NATURAL DISASTER

I COVER THEM. I AM ONE.

GINGER ZEE

HYPERION AVENUE

LOS ANGELES • NEW YORK

This book is dedicated to my late grandparents,
Oma & Opa (Hilda & Adrian) Zuidgeest, George Hemleb,
Paula Wesner, and John Craft—and to my beautiful grandma
Clara Craft, who remains a wonderful part of my life.
Without all of you there is no me—no mess—naturally.

All rights reserved. Published by Hyperion Avenue, an imprint of Buena Vista Books, Inc.
No part of this book may be reproduced or transmitted in any form or by any means,
electronic or mechanical, including photocopying, recording, or by any information storage
and retrieval system, without written permission from the publisher. For information address
Hyperion Avenue, 77 West 66th Street, New York, New York 10023.

First Hardcover Edition, December 2017
First Paperback Edition, October 2021
10 9 8 7 6 5 4 3 2 1
FAC-025438-21239
Printed in the United States of America

This book is set in Goudy Old Style
Designed by Julie Rose

Library of Congress Control Number for Hardcover: 2017018075

ISBN 978-1-368-07651-7

Visit ginger-zee.com

CONTENTS

INTRODUCTION

Ten days before I started my job at ABC News, I checked myself into a mental health hospital. This is a compilation of stories leading up to and after that fateful decision that I hope will help demystify depression. This is the anti-Instagram, the raw, sometimes scary, and hopefully humorous life I have led so far.

This is me, a Natural Disaster.

I cover natural disasters, and I've struggled with being one in my personal life.

From Hurricane Katrina to California wildfires and blizzards in Boston, I've covered the nation's largest natural disasters for more than a decade. I'm intimate with human reaction and interaction during Mother Nature's greatest fury. Coincidentally, through that process, I have finally started to figure myself out and oddly found quite a few parallels.

Earth is just one big ball of energy that is constantly attempting to find balance. The poles are cold and the equator is hot. Earth wants so badly to equalize that difference. In doing so, it creates storms. Many of them are dangerous. But all storms are necessary.

This helped me realize that all I am is just one big ball of energy that will at no point be constant, even though that is what I constantly yearn for. We get average temperature or average precipitation only from having extremes. That's what life is: a series of extremes. The sooner we can embrace that, the more peaceful I believe we can be.

Very few natural disasters catch us by surprise. We see them coming, whether it's through technology or our own eyes. And with that time, we make choices that can determine our fate. Do we stay in our homes or decide to flee? Do we shelter in the basement or go outside to watch? And then there's the aftermath. Do we let the natural disasters destroy us and consider ourselves victims of forces beyond our control, or do we find gratitude in being alive and being stronger for the experience? And most importantly, can we shift our focus off our own perceived tragedies and reach out a hand to somebody else going through the same type of storm? That's why the title of my book is so meaningful for me. My job in some ways has helped me get through my own personal storms just by seeing the disasters as a metaphor for a universal human experience and has given me invaluable perspective along the way.

When I witness a natural disaster, I am particularly struck by the degree to which we are all the same. Natural disasters leave tremendous shock, destruction, and sadness in their wake. My biggest regret as a natural disaster is that I did, too. Most people walk around after the initial tragedy, bewildered. Then they often start acting irrational, looking for house keys to a home that is no longer standing, or standing in line at a drugstore that's been knocked to the ground. Eventually, sadness and some level of acceptance sets in. Hopefully, in the final phases of grief, we realize we are grateful to be alive and that we need to shift our focus off ourselves and onto helping others. We've all seen the footage and pictures of the first responders and the ordinary people who put themselves at great risk to make a tourniquet out of a belt for a victim or swim into dangerous waters to bring a stranded boy and his dog onto a boat. It's quite inspiring, and in a weird way it's what convinced me to write this book.

Initially, I flinched at the idea of writing a "memoir," because I don't think anything I've done in my life makes me that important. But then I thought of all the natural disasters in my own personal life that I have survived, how I've grown stronger from them, and how I'd like to share those if there's any possibility it will bring hope to somebody in the eye wall of their personal hurricane or comfort to other survivors to know they are not alone. No matter your storm, it never rains forever. It can't and it won't.

That's where I am now. By no means do I think that I've figured life out or that I'm some model of perfection—but it's not raining. In fact, it's pretty sunny on my side of the street. I'm happily married and the mother of a beautiful baby boy, and I have my dream job. And while I still have a lot of road left ahead of me, I'm at the point where I can look back at all the so-called disasters of my life—the sometimes chaotic childhood, the failed engagement, the insecurities and people pleasing, and the abusive relationships, and just like when I've been lucky enough to watch a tornado approach, I can find the beauty and strength that was born from each of my "storms."

I can now accept the here and now. I understand that life isn't always full of sunshine. There are many rainy days. Clouds that persist for years at times. And even at this point in my life, where everything is quite bright, I know it won't be like this forever.

I also know that because of the chaos I created, the natural disaster I used to be, I was forced to do the kind of deep, soul-searching growth that means there will never be a next time when I find myself crying and drunk under a bridge in Chicago. There will never be another time when I'm hiding under a hotel table calling the police because my boyfriend emotionally abused me.

I hope you get a good laugh—even at the sad parts, I do. Finally telling these stories feels good. I think even some of the people closest to me may not know all these details. But this is me. A *natural disaster*. And this is me finally being okay with admitting that.

Chapter One
RUNAWAY BRIDE

I canceled my first wedding . . . twice. Which is why I will *always* watch *Runaway Bride* when it comes on cable. I'm not ashamed to admit that it makes me feel good about myself. Julia Roberts ran away *four* times from *four different men*, one of them being *Richard Gere*. My runaway technically makes me at least 50 percent more efficient and way less erratic than the goddess of romantic comedies. Unfortunately, at the time, I did not possess the kind of clearheaded confidence that would have allowed me to listen to my instincts and save everybody a whole lot of drama, pain, and security deposits by canceling my wedding just once. The fact that my fiancé, Joe, was a great guy, the kind of guy who fixes things and plays catch with the neighborhood kids—plus lets you watch *Sex and the City* reruns even during March Madness—made me think I was nuts.

On the other hand I wanted to be a meteorologist since I was a nine-year-old staring at the storms rolling in off Lake

Michigan. I knew that the life I had seen for myself for close to twenty years—chasing tornadoes, jumping out of planes, swimming with sharks—probably wasn't going to make me your typical mom, which is what I thought Joe wanted and needed. He was a solid guy, simple and sweet. He deserved the same. But that isn't me, and it took a lot of time before I could be honest enough with myself to realize it. I also never gave him a chance to express whether that would or would not work for him or for us; instead I avoided confrontation and just let the engagement go on.

Joe and I were engaged for eighteen months—a long time by most standards—and that's on me. We got engaged after only six months of dating, and we knew we needed some time before the wedding. I feel it's important to mention this because there's an idea in our culture that it's the guy who drags out the engagement because he's not dying to say, "I do." But this was a joint decision, probably driven more by me. I'm not saying this is cool or that women should aspire to an inability to commit; I'm just saying women should be allowed to be as messed up and complicated as men. In fact, I don't have one female friend who hasn't had a time in her life when the last thing she wanted was to settle down. That's why it's a lot easier for a lot of us to date jerks. Somehow the problem resolves itself, and you (and by *you*, I mean *me*) don't have to look in the mirror too long.

It's just so much harder to break it off with a nice guy. It's not just that you feel like a horrible person for hurting somebody so kind; you spend a lot of nights staring at the ceiling convinced you must be losing your mind, there's something wrong with you, and you will definitely regret this one day. And this sucks. At least it did for me.

In my opinion, too often women are told that our ambivalence toward marriage is something we must work through, in therapy if necessary, so we don't get left behind and spend the rest of our lives with a bunch of cats and a freezer full of Lean Cuisine. The truth is, we're complicated, and learning to trust ourselves is so much more important than the pursuit of marriage. It took a while to learn to trust that Oprah voice that always told me to live my best life and embrace my "aha" moment. And that meant breaking up with Joe. Twice. Even after the invitations were mailed and the cake was ordered and—for the second cancellation—the guests were on their way.

As my wedding to Joe drew near, I was working as a reporter and meteorologist at WOOD TV in my hometown of Grand Rapids, Michigan. Even though I was on television, being a reporter was not fun for me, because all I'd ever wanted was to be a meteorologist. Besides, that's what my contract said I was hired to do—report on the weather. Unfortunately, when I started at WOOD TV, all five of the meteorologists already on staff suddenly decided they had no intention of retiring or moving on as the news director who had hired me had anticipated. And since the station needed a reporter, and I was under contract, that's what I was told to do. And I did it to the best of my ability. I wasn't exactly miserable, but I wasn't exactly happy, either. This is important as context for "the wedding that wasn't," because my restlessness at my job was feeding the flames of the restlessness in my relationship.

Now, please don't misunderstand all this grown-up self-awareness to be anything but hindsight. Which is just to say, I am, as always, aware that I am a natural disaster, softened these days by my deep regret that I ever caused anybody in my path any grief or pain. Especially Joe.

Joe and I were in love, and in Grand Rapids, there's a clear path: go to college; find a decent man; marry and have kids with him. And besides, he asked, and I have a hard time saying no. In fact, my friends and family say I can't spell *no*, which is ridiculous. Of course I can.

Even today, as a big-ass grown-up mom/wife/career person, I'm still working on saying no when it's appropriate. I feel like I just came into this world ready to say yes. People like to talk about being an old soul; well, I've always felt like a really new soul, like maybe it's my first time here, so I want to try everything. That's the kind of fearlessness that has unconsciously become my brand, and I love it. If a producer pitches me a story idea, I always say yes. The wilder the better. And while this is a great quality for my work life, it is maybe not so great for my personal life. The way I see it, if you jump out of a plane or paraglide through the Andes, the worst thing that happens is you die an instant death. But I'm a statistics gal, and statistically speaking, those adventures will not lead you to your death. But if you marry too young or marry the wrong guy, statistics are not kind. It can be the kind of slow death one might suffer if forced to ride a carousel endlessly. Pleasant and safe, but eventually a little dizzying and suffocating. This metaphor should give you a pretty good idea of where my head was at exactly six weeks and one day before my wedding to Joe.

There is standard etiquette in the wedding world that you have to send your invitations out at least six weeks before the big day, so of course I waited six weeks and one day before the wedding to send out mine. And less than twelve hours after they were dropped into the mailbox at my neighborhood post office, I woke up in a cold sweat. Let me just say that I hope menopause never brings me the kind of sweat I had that night, because it was awful. Making the night even more entertaining was an adrenaline rush that fight-or-flighted me into action. Suddenly, no more fear. *Poof!* It was gone. I snuck out of bed like a cat burglar, even though Joe always slept like a bear in the middle of winter. I pulled on some running sneakers and ran out the door.

It was still dark outside, and the local post office was three miles away, but I had no trouble running there, crying with every step. I had no plan, just a vague conviction that those invitations could not under any circumstance depart from that mailbox. After about ten minutes of wiping the snot from crying on the sleeve of my pajama top, a postman tapped me on the shoulder.

"Let me guess," he said. "Sender's remorse?"

"Yes," I said. More snot, more wiping.

He took out a ring of keys from his blue pants and began to unlock the box. I began to hear angels sing. He was going to help me!

"This happens all the time," he said, with a light chuckle like I'd just done something cute like dropped my ice cream on the floor.

I didn't care what his point of view was on this matter, just that he was going to help. It did occur to me that perhaps the US Postal Service needed to vet their postmen a little more carefully if it was this easy to get them to commit mail fraud. But obviously I kept my mouth shut. The last thing I needed was for my accessory to come to his senses.

It wasn't hard to find the one hundred oversized silver envelopes in the midst of the utility bills and birthday cards. I grabbed the envelopes in snotty fistfuls, and my postman helped. Clearly, and for whatever reason, he was all in. And when we had them all and he'd locked up the mailbox, he put his hand on my shoulder.

"Now you get going, miss. I have a feeling there's someone at home you need to have a talk with."

He was right. And in that moment, I wanted to marry *this* man for being so kind and helpful. I went in for a hug to thank him. He pulled back. The crime was over and he was moving on. Understood. Respect. I nodded and turned to go. I took my time going home, exhausted less by what had just happened and more by what was going to happen.

A few hours later, I approached Joe with the box of invitations, organized as if they had magically never made it into the mailbox. But he knew they had—we had done it together less than twenty-four hours before. He was understandably confused. In a few garbled crying words, I simply told him the truth: that I couldn't marry him.

I remember his reaction as if it was playing out in this moment. Joe somehow managed to move through all the Kübler-Ross stages of grief in less than thirty minutes. First he threw a laundry basket, which was so out of character (anger). Then he started saying we could go to counseling together (bargaining). Then he began to weep (depression). Then he realized that if I wasn't ready, then we weren't ready, and the wedding just wasn't going to happen (acceptance). And finally, we landed on the big one. Denial.

"It's cold feet," he said. "It happens to everybody! Maybe you should talk to somebody. A pastor, your mother, a therapist. I'll talk to all of them if you want. Just trust me, this will pass."

And then he began to cry again. And I began to weep.

And then the wedding was back on. I didn't want to hurt him, and my instinct is always to do whatever is necessary to stop anybody from crying. Staying the course and trusting my instincts by canceling the wedding was just too hard. Lots of people besides Joe would be mad at me. My beautiful vanilla buttercream cake would never be eaten, deposits would be lost, plane tickets canceled, dresses returned. Joe's version, where my anxiety was typical and would eventually pass, was way more manageable. All I had to do was make it to my wedding day, and I already had a strategy that seemed pretty darned clever.

For three weeks I subsisted on a rabbit's diet of no more than a fistful of carrots each day. When you are very, very hungry, your whole brain focuses on that and it's impossible to have any other problems, because it's all about the carrots. As someone with years of an eating disorder under her belt, I can tell you that hunger can be powerful. Hunger has to be a big reason why cavemen did not need psychologists.

And then, about three weeks before the wedding, I went cherry picking with my *oma* (Dutch for grandma). Oma's name was Hilda. She had moved here from Rotterdam in the Netherlands with my *Opa*, Adrian, and their son—my dad—in the late 1950s. Oma was a tall, hilarious woman who was known for her bluntness. When her weathered hands had been picking cherries for more than an hour, she shot me a look and popped a cherry in her mouth. Through her thick accent she said, "Sit with me."

We had a deadline for cherry picking, because we were going to use the cherries to make jam as wedding favors, and imagining Martha Stewart's approval brought me comfort in the midst of all this chaos. Oma had grabbed my basket when she saw I was concerned about the time. With her beautiful piercing blue eyes, she then shot me a second very serious look.

"If you don't want to get married, you don't have to."

I froze. It was like she'd thrown a rock and hit my head with that one sentence. I felt light-headed, and not just from hunger. This woman, who had been through a war and traveled to America to start her life over with nothing but a trunk and a few words of English scribbled on a piece of paper, had always been a voice that demanded my attention and respect. And here she was, offering me permission to say no. Past the etiquette deadline. Past the time when any of my guests could get their tickets refunded or cancel their bridesmaids' dresses (although in my opinion they were a vibrant shade of green that could have easily been transformed into a lovely summer tablecloth down the road). Past the conversation where I'd agreed with Joe that my USPS raid at dawn had been nothing but normal wedding jitters. But most of all, she was giving me the gift of seeing that it was okay to listen to the voice in my head that I couldn't allow to be heard.

In retrospect, that is the essence of my depression. I often knew when something I was doing was wrong, but I rarely had the confidence to make the difficult choices—always avoiding confrontation and allowing the hurt and pain to stay inside me. It sounds like you're being protective of others, but in truth it's a cowardly way of living, avoiding difficult communication so you can get where you want to be. In this case the gravity of my decision would have legal implications and it is one of the first times I let the right voice override the invertebrate that typically resided in my body.

My mom joined us and supported my decision wholeheartedly. If I'd had the strength and hadn't been barely surviving on war rations, I would have skipped through the fields. But I needed whatever strength I had left to face Joe with the truth and to stick to it.

I don't know if I blocked it out, but I honestly don't really remember the details of how I finally ended it with Joe. I do remember that this time Joe knew I was serious, and he was devastated. I knew that he really loved me, that he accepted my crazy unconditionally. I wondered why that wasn't enough, but I did not waver. There was no turning back. He pleaded with me to go to counseling and postpone the wedding until I was ready. The yes-girl inside me who people pleases as naturally as she breathes entertained that idea for a moment. It would be so easy. Just forget about my needs and my instincts and marry this guy. Luckily, the woman my Oma recognized in the fields put her foot down and said, *No. I can't marry you. If it's not right now, it won't be right later.*

I enlisted support so I would remain firm in my decision. My mother and my Oma had my back, as did my stepfather, my dad, and my brother, and as each day passed (and I began to up my calorie intake), I grew a little bit more confident in my decision and prouder of myself for making it. I was telling the truth. My truth. Even though I owned a house and had a real job, canceling my wedding was the first decision of my relatively young life that made me feel like a grown-up. And even though it took me two tries, I was finally making a very difficult decision that would ultimately and undoubtedly save both Joe and me a lot of misery down the road. I knew I was jumping into the unknown, away from the life I thought I was supposed to have, and somehow I just knew that it would all be okay.

My mother loves to tell the story of how, when I was eighteen months old, I stood at the top of the stairs, looked her straight in the eye, and jumped. I can only imagine the fear, wondering how she would survive the next eighteen or so years with me as her daughter. And whatever she thought the future might hold, she probably wasn't far off. I was the high school cheerleader who thought the top of the pyramid was the only worthwhile spot (although I was almost always a base because of my weight and height, and ironically needed more stitches than the cheerleader at top).

In college, I started storm chasing, definitely not a mother's dream for her daughter. And now, I'm a meteorologist, a job that certainly does not require being a daredevil, but that's the part I love the most and have fought to work into my career. In fact, if I'm not taking risks I'm miserable. Even when I'm on television just talking, I don't use a script; it's all ad-libbed, and I am rarely scared in any of the adventures I take on.

In the end, my non-wedding became an impromptu family reunion with double the champagne and cherry-jam party favors. It felt odd to be partying at all given the heartache I had caused and the heartache I was feeling, but you could say we were all celebrating how Ginger finally learned to say no.

It turned out there was another quite unexpected reason to celebrate. The day before my non-wedding, I got a call from the news director at WMAQ in Chicago that the weekend meteorologist job I'd interviewed for months ago was mine. It was a network affiliate in a major market, and I was over the moon. I would be leaving behind the reporting job and the memories of Joe and starting a brand-new adventure. I felt like I'd jumped and landed more than safely. I felt like I'd finally landed in my life.

Chapter Two
A WEIRD LITTLE GIRL WITH EVEN WEIRDER DREAMS

There are really only two places to watch the beautiful formation of a growing thunderstorm: the vast open fields of the Midwest plains, and over any large body of water. It was Lake Michigan where I first fell in love with the weather. The course of my life was set at age nine. I consider myself very lucky that I found my passion so early, and also that my parents got divorced.

I know that sounds crass and completely random, but it's not. I really don't believe I would have found my passion as early or as intensely had my parents stayed married. Because if they hadn't gotten divorced, my mom would not have been (briefly) engaged to a man whose cottage on Lake Michigan sparked my enthusiasm for the weather.

My dad wasn't a big fan of the guy and nicknamed him "Dickhead," which my brother Sean and I thought was hysterical. My dad allowed Sean and me, five and eight at the time, to call him Dickhead, and that was just about as much fun as watching *Beavis and Butt-Head* or *In Living Color* (which was also allowed only by my dad). My dad is a handsome, hardworking, quiet guy who emigrated with his parents from Holland when he was six years old. Almost immediately, my oma and opa moved to Western Michigan, where thousands of other Dutch folks had settled. He grew up loving sports. Reading wasn't easy since English was his second language. Still, he graduated college with a degree in geology and started traveling with his high school buddy to see more of the United States. They basically worked to ski and enjoy life, though on one of their many stops they briefly worked for a tennis-court contractor in California. And with that entrepreneurial streak so common in immigrants, he and his buddy took that skill back to Western Michigan. While they were plotting the start of what would be a successful thirty-plus-year business building tennis courts, my dad was making money as a gas station attendant. One fateful day, he pumped my mom's gas. I know, that sounds so double entendre-ish and cheesy, but it's true. My mom, a beauty from Long Island, New York, had moved to Michigan for the diving and swimming team at Michigan State University. She had studied nursing and was currently teaching in Western Michigan, and she needed her tank filled.

They dated for a year and then broke up. During that break my mom realized she was pregnant. They really weren't *together* together, but my mom showed up to my dad's softball game and said, "I need to talk to you. I'm pregnant." And my dad hit a home run. Or so the story goes. They married in Grand Rapids, and my mom still tries to say you can't see me in her wedding dress. You totally can. The Empire waist gives this almost-bastard child away immediately. Before I was born, they moved to California so my mom could get another master's degree. (My mom *loves* school.) Eventually we moved back to Western Michigan, they had my brother Sean, my dad worked hard, my mom worked hard, and they just did not work together. And so they divorced. Enter Dickhead.

Dickhead had a cottage on Lake Michigan. Like, *on* Lake Michigan. It was one of those pricey joints we had never even visited, and for Sean and me, the house almost made up for the guy being such a jerk. Well, we might not have known the full extent of his jerkiness quite yet, but in my dad's eyes, he was the first guy my mom dated after they broke up, so that was jerky enough. He would graduate to magna cum laude jerk soon, though. The lake house Dickhead owned was a huge gray modern design that opened up to a deck where it felt like you could see forever. A perfect heath covered in tall grass spilled up to a wide-open beach that was all our own. I remember countless afternoons when the storms would start gathering on the lake. We would hear my mom start freaking out. (She always loses her mind in storms, which is pretty ironic given that she gave birth to me. Her threshold for a freak-out is about a twenty-mile-per-hour gust, and then she yells *"Everyone down in the basement!"* at the top of her lungs.) I loved watching the base of those thunderstorms, the billowing tops of the cumulonimbus, the lightning that effortlessly lit up the lake and sky. It was gorgeous, so energetic. I was in love.

Before we ran inside, I would wash my feet as fast as I could in the outdoor shower. I wanted to get up and watch the storm coverage on television with my mom. Were these storms dangerous? What would they do to us? Why was my mom acting this way? I watched the meteorologists on my local station showing us the warnings. It was so fascinating to me that those meteorologists became my idols. I was like a kid from Cleveland who wanted to grow up to be LeBron James. These scientists on television inspired me to be a meteorologist from an early age, but I never thought I would be on television. Actually, very few meteorologists are on television. Most meteorologists work either for other private companies (think insurance, trucking, and shipping) or the government.

The rest of that summer played out with many more storms and culminated in a true disaster. Dickhead left my mom six days before their wedding. Now, my mom is one of the most passionate women I know: strong and bulldog-ish when you need her to be, scary at times because of her intensity, but soft and fragile, too. My mother is the ultimate caregiver, but when she reaches her limit and doesn't feel people respect her or value her efforts, look the F out. I've seen my mom fly off the handle and cry more times than I care to remember. But this time was different. I woke up to the sound of deep, guttural cries. My aunts and grandma were there consoling her. At nine years old, I didn't understand exactly what was going on, but I do think I started to feel better about calling him Dickhead.

Dickhead had gone to my dad's house and asked my dad to take my mom back. He said he couldn't handle us all. He gave up on her, and us, days before we were supposed to become a family.

My mom was understandably devastated and embarrassed. Ironically, much like my non-wedding years later, we still had a party with all the goods that were meant to be at the end of the aisle for my mom and Dickhead. Our family gathered to celebrate being together. It was a family reunion in place of the union that was supposed to take place. What a mess, but a beautiful mess. That's really the point of this story: while my mom was and always will be a diva of disarray, through perseverance and good-heartedness, she's been able to come out of everything with the most beautiful life. She has four great children and a twenty-five-plus-year marriage that's going strong, all of which came from her ability to say yes and allow life to happen in good and bad times. I know which side I get most of my natural-disaster traits from (hers), and this is also one of the first points in life where I remember finding some pride in it, too.

Oh, and to you, Dickhead, if for any reason you or anyone related to you is reading this, I want to say thank you. I am so grateful you didn't marry my mom, because that meant we got my stepdad, Carl, who is the antithesis of a dickhead—and I still fell in love with weather on your beach and got to follow this unbelievable path. With maturity and a failed engagement of my own, I commend you for making the best choice for everyone, albeit a difficult one.

Throughout school, science and math came most easily to me. Since that summer on Lake Michigan, I had it in the back of my head that I wanted to be a meteorologist, but at first, I fought it. It seemed so far-fetched.

In my senior year of high school, I took a class called TV Studio, which was taught by a former television reporter named Colleen Pierson. She was an encouraging, sweet woman who embodied a 1990s reporter/anchor with her bright pantsuits and even brighter coordinating lipstick. Colleen gave me a fateful push by suggesting I should try TV weather forecasting as a career. At the time, it seemed ridiculous, and I dismissed it. But I stayed focused on meteorology.

I chose my college, Valparaiso University, not only because

I could get a bachelor of science in meteorology, but because they actually had a class called Storm Chasing. I had seen the movie *Twister* when I was in high school, and the dramatic storm-chasing scenes in the movie struck a chord somewhere deep inside of me. If I could figure out a way to do that when I grew up, I would be happy. But I didn't just want to chase the tornado for the adventure; I wanted to gather the data like Helen Hunt's character did, dissecting the atmosphere and making the calls. That was enticing.

So I dove into my meteorological studies once I arrived at Valparaiso University in northwestern Indiana. The college is in the small town of Valparaiso, about an hour east of Chicago and about two and a half hours south of my home in Grand Rapids. During my freshman year, one of my professors, a lovely man named John Knox, suggested I do an internship in television. Now I had Mrs. Pierson and Professor Knox telling me they thought I should try television meteorology. My confidence and interest growing, I sent about three dozen résumés out to television stations, mostly within driving distance, and heard nothing back. Not one response. I guess nobody wanted a freshman as an intern. But just as I was about to leave for a summer of bartending and hanging with my high school friends back in Western Michigan, I got a call on my dorm-room phone (note, we did not yet have cell phones) from a woman with a very thick southern accent.

"Y'all know Birmingham?"

The answer was no. I was a Michigan gal currently going to school in northern Indiana. I had never been to Alabama. But I lied and said, "Sure do!" A week later I moved to Birmingham for the summer. It was fate. And one of those moments when my natural-disaster spontaneity worked out.

My internship was with James Spann, a weather king in our industry whom I credit with solidifying my interest in pursuing a career in television meteorology. Within the first week of the internship, James invited me to accompany him to an elementary school that had invited him to talk to the students. He got up in front of about a hundred children, and he might as well have been Justin Bieber during the peak of "Baby" at a mall full of twelve-year-old girls. But he wasn't singing pop music. He was singing my tune: science. I saw each of those adoring faces light up, nod, and get excited about the weather because of James's passion. It was infectious. I saw in that moment what has come to be my favorite part of my job—the communication of science.

Three days a week I would show up at ABC 33/40 for the five, six, and ten P.M. shows, dripping in sweat (my car didn't have a great air conditioner, and summer is no joke in Birmingham). Usually I would be coming from my money-making job (this was a nonpaying internship) as a beer-cart driver/pool attendant at a country club. From that extreme heat, I would enter the refrigerator that was the studio. I have worked in at least a dozen studios now, and they are all way too cold. When I have asked, they always say it's better for the cameras, or the monitors, or whatever electronic device happens to be in my line of sight. I just don't buy it. I think it has to be the men in suits setting the controls, because my television monitor at home seems to work just fine at seventy-eight degrees.

Frozen or not, I loved every minute of the time I spent with James. I got to see the behind-the-scenes workings of a television meteorologist. It was the school talks, the radio shows, the family James doted on, and the community he was a huge part of that I wanted. I still have never seen anyone do it better than James.

When I returned from that internship, I became focused on getting a job in television. And on the very first day of my sophomore year, I ran into a poster hanging in the meteorology department that read ON-AIR TV METEOROLOGIST WANTED. I couldn't believe it! The ad was from a PBS station called WYIN. I looked around, hoping I was the only one who had seen it. Most of the students in my meteorology class wanted nothing to do with television, so I didn't have much competition there. But I heard there were several upperclassmen who had seen the same ad and were going to audition. I drove the twenty-five minutes to the small town of Merrillville, Indiana, where the station was based. I was so nervous. Was I really going to audition? Against juniors and seniors? I was only nineteen! I walked through those doors with a "What's the worst thing that can happen?" attitude, and amazingly, I came out with a job.

At WYIN I went by Ginger Zuidgeest. Meteorologist Ginger Zuidgeest. That is a mouthful, and the male anchor who had been there for at least thirty-five years always stumbled a bit on my last name, but I didn't care. I was on TV! The viewers probably numbered about a dozen, and the job didn't pay, but my career was starting, and it was an exhilarating time. I would show up around three in the afternoon and tape my weather segment. We actually stole The Weather Channel's graphics from their website and photoshopped their logo out. So strange, but I guess at PBS you can get away with a lot (who's going to pick on a PBS station, right?). I had a little desk in the corner of the cold studio, but it was all mine. The set was basically one old camera, a news desk, and a green screen. But that green screen was the first that I ever had regular practice in front of. I learned a lot in front of that green screen, like where to stand and how the screen acts like a mirror so your movement is actually reversed in the camera when you see yourself. It's where I became comfortable looking into a lens. I wish I had some of those tapes, but I saved nothing.

I kept that job the rest of my time in college, cohosted a

radio show, and took on two more internships before graduating with a bachelor of science in meteorology and minors in math and Spanish, and I did it in three and a half years. At this point in my life, I had all the confidence in the world. Looking back, I may have had a slightly natural-disaster inflated sense of my accomplishments. I felt like I had so much experience and what I thought was a pretty awesome demo reel. So I went out and bought ninety-nine VHS tapes, each carefully labeled with what I believed to be my best-looking head shot, next to my name, e-mail address, and phone number, and I sent them out and waited for the network calls to come flooding in.

But all I got was silence. *No* job offers. Zero. Zip. Nada. Crickets, as they say.

I was humbled, but I knew my destiny could not be denied. So I did what any aspiring TV anchor with no prospects does: I took a job in radio and bartended on the side to pay the bills. That radio gig was a favor paid to my guardian angel, Ed Fernandez. Ed was the general manager at WXMI, where I had had one of my internships. He said he saw something in me and wanted to make sure I got into media. He introduced me to the radio legends who worked at WLAV. WLAV was a classic rock radio station in Grand Rapids where the other hosts were nearly three times my age, and I loved it. I felt like I was "cutting my teeth" in the real world. I learned how to tell a weather story in fifteen seconds without pictures, and to do it in front of gruff classic-rock dudes who must have been so perplexed as to how or why I was sharing their booth. It was that booth where Ginger Zuidgeest became Ginger Z.

Zuidgeest is a long, extremely Dutch last name. And as much as I love my dad and our name, Tony Gates, Uncle Buck, and the crew at WLAV told me, "You only get fifteen seconds. We aren't going to waste half of it on saying your name." And that is how Ginger Z was born. A few years later, I would tack on the two Es to make it sound more legit (thanks, Peter Chan), and it's worked out pretty nicely. Nobody ever says, "Hey, Ginger!" They always say, "Hey, Ginger Zee!" It has a lot of bounce to it, I guess. In fact, several viewers have named their dogs Ginger Zee after me.

While I worked at the radio station, I kept sending tapes out to every job opening I saw. I sent tapes to Lima, Ohio; Fargo, North Dakota; and even Flint, Michigan, the home of Michael Moore's *Roger & Me* and unfortunately the lead-pipe contamination capital of America. And at the age of twenty-two, my dream came true when I was hired as a full-fledged meteorologist at WEYI in Flint.

Chapter Three
THE FIRST HINT
OF DEPRESSION

In the six months after graduation and before I got the call for Flint, I encountered my first real bout with depression. For anyone that has ever struggled through the hopelessness that defines most depression, you know that transitions can be a major trigger. That period after college and before I started my job felt (as it is for so many young people) like a loss of control, a loss of faith in my talent, and for me, the first time I remember feeling low. Not like, "oh I'm having a bad day," or "I'm in a funk-low," but the kind that keeps you from friends and family, chains you to your bed, and had me thinking food was unnecessary.

During my final semester of college in the fall of 2002, I was traveling back and forth to Valpo one day a week to finish a class there while living and working in Grand Rapids. I would often find myself falling asleep in class and even falling asleep driving. For a type A person, falling asleep was just about the most disappointing thing I could do. I had too much to accomplish to sleep this much. And even when I slept, I felt like I was never getting rest. I would get so angry with myself for not meeting my insurmountable goals.

My alarm would go off at 4 A.M. I wasn't yet on a morning-show schedule, but rather, I had filled my Valparaiso University planner hour by hour.

4 A.M.	Wake-up, sit-ups, push-ups
4:30 A.M.	Bed and breakfast work
5–9 A.M.	Work
10–11:30 A.M.	Class
11:30 A.M.	Lunch
12:15–3 P.M.	Class
3–5 P.M.	Run, weight workout
5–6:30 P.M.	Shower, study
6:30 P.M.	Dinner
7–10 P.M.	Study, push-ups, sit-ups, bed

Every day was filled, every moment planned. When something got in the way of me making money, or more importantly, me working out for at least two hours, I would be furious. I constantly told myself I didn't have time to waste. Especially on sleep, or car accidents.

Fast-forward to one bad car accident later where I almost crossed the median into oncoming traffic on a highway; I knew I needed to get help. I thought I had a heart problem so I went to a cardiologist. Several tests later they assured me I was in tip-top physical shape. They referred me to a neurologist who suggested we do a sleep study. With all the cords attached from scalp to ankles, I fell asleep for eight hours. They then woke me and I was told to take four half-hour naps every two hours. I fell asleep within minutes of each nap and they didn't even need to finish the testing before they told me I was narcoleptic. I was prescribed a drug called Provigil and suddenly the world became clear. I always describe my first day taking Provigil like the scene in *The Wizard of Oz* when everything turns to color, or when the black bars on a movie are lifted and the picture fills the screen. It isn't a stimulant, but an alertness medication I was told. I could smell, see, hear, and most significantly, I could feel.

I often think that narcolepsy was God's way of protecting me from myself. I was such a serial perfectionist that the dullness I felt as a narcoleptic in my later teen years (when the disease usually develops) probably protected me from self-harm. As pleased as I was to be on the drug that allowed me to finally feel alive, there were consequences. All of my feelings were amplified. My highs were even higher and my lows felt as deep as the Mariana Trench.

Wallowing in my sea of rejection when no one had responded to my résumé tapes, I filled my newfound awake time juggling work, friends, and boyfriends.

I had been a serial dater from the age of fifteen beginning with my high school boyfriend. He was a terrific guy, an awesome influence (we didn't drink because we were both in sports and made a pact with each other). He tucked me in every night, rubbed my head and feet, and called me Princess. He and I stayed together through the start of my sophomore year in college. I think we both believed that we would be together forever, giving in daily to the delusion and pressure that surrounds the phrase "high school sweethearts." We didn't make it forever, of course. But the attachment I had to him kept me always looking for someone to be by my side. Someone that loved me like he did. My Aunt Darci explained it to me once as "my imprint." At that crucial time when we are developing into women, the man, or dating experience we have, makes an imprint on your life. He was my imprint, and from that moment on I was always in search of that fairy tale he and I had drawn up.

I had an insane fear of being alone. I found such comfort in having a boyfriend; they validated me and bolstered my suffering self-esteem. I needed them. Or so I thought.

When the career radio silence combined with the end of my relationships, I moved home single and feeling hopeless, the life I had dreamed for myself shattered. I was supposed to be this big time TV meteorologist with a handsome guy who would likely become my husband so I could soon have a family. That was the type of lofty expectation, of perfection, that drove me but also caused major harm.

When the going got tough, I broke. I believe my new drug enhanced the *low* feeling, making me feel like life was not worth living.

One night I snapped. My high school sweetheart was back and wanted to try being together again. He had come over to discuss this possibility and I couldn't be honest with him. I still hadn't learned how to say no and tell people how I really felt. I felt trapped by my inability to speak my mind. I didn't want to upset him and I didn't want to lose him forever. Suddenly there was this loud voice screaming at me, telling me that I was not strong enough, not good enough, and assuring me that my life was worthless. I was mentally transported to a strange dark room I had never been in. I felt imprisoned there. I couldn't see light anywhere. It was cold and damp and no one could get in; yet oddly I didn't want out. This is what depression looked and felt like to me. While all of this mental bending was going on, I listened to that voice. I locked myself in my bathroom and took out every bottle in my medicine cabinet and consumed every last pill.

Thankfully my roommate and ex-boyfriend begged me to come out of the bathroom; they saw that I had taken everything and called 9-1-1. Thankfully the concoction I had taken was not lethal, and at the hospital they told me how lucky I was. "Had you taken this quantity of acetaminophen, this would have been a different story. Thank goodness it was mostly Benadryl and other relatively benign substances. Either way, I believe you need some help."

They suggested counseling and I went. I eventually ended up going to counseling a lot, but I don't think I ever went there ready to listen or to enact change.

My mom took me back to our house after my suicide attempt and insisted I stay with her until she felt I was safe. When I went to the bathroom, she asked me to keep the door open. It was so odd when I looked at myself in the mirror. I didn't see the girl who had taken all those pills. I saw me again—as if it never happened. I wasn't loving life by any means, but I couldn't believe a day before I had tried to take my own life. That's one of the wildest parts of depression. It can be so fleeting. Fickle, really. And very difficult to plan for. In the years to come, I would pop in and out of that dark room effortlessly. I would be having a fabulous day, whistling Dixie, and all of a sudden the windows would shut and the darkness would settle in like a fast rolling fog, obscuring all the light. The dark room usually came when I was confronted with conflict, making difficult decisions, or going through transitions. And just as quickly as it had settled in, I had an uncanny ability to pretend that dark room never existed as soon as the lights flickered back on.

We have these scary bugs in our new house called sprickets (a spider and a cricket combined). They remind me of depression. They live in dark, damp places and jump at you when you least expect it. Unlike the spricket, which can't really hurt you, depression can.

Soon after the suicide attempt, I got the call from Flint, and for a short time my life had purpose again. I left behind the memory of the desolate girl in the cellar without windows and moved on—forgetting any of it happened, focused on the future. But it, my dark room and that other version of me, would return—again, and again, as sure the seasons.

Chapter Four
OTIS/FLINT

Every natural disaster needs a sidekick. Mine was Otis. Otis was a black lab who passed away just two weeks after my son was born, only a few weeks short of his own fourteenth birthday in early 2016. This dog was with me through every job, every city, every boyfriend, every apartment, and every *natural disaster*.

I got Otis by sheer coincidence. A friend had heard from a friend that a dog breeder had two puppies left over from a litter that were not going to be showable because there was "something wrong with them." But when I went to meet Otis, there was nothing wrong with him. It was love at first sight. That dog was meant for me. He whimpered as we drove home from a farm in Holland, Michigan. If I hadn't taken him, the breeders were getting ready to drop him off at the Humane Society, which wasn't a terrible place, but it wasn't home with me.

Within days, Otis's unique personality started emerging. For instance, there was the time I woke up to the sound of an entire forty-pound bag of dog food cascading into his crate. Otis had gotten just tall enough in one week to nibble at the bag, which I had carefully placed on top of his crate away from him—or so I thought. The nibble opened a hole that kept getting bigger under the weight of the bag, and this waterfall of kibble was my fat puppy's dream.

The look Otis gave me when I busted him for his kibble hijacking was a lot like the look my son gives me when he gets caught—equal parts guilt and charm. This dog had a twinkle in his eye, and if dogs could have dimples, Otis did. He stole my heart in every moment we got to spend together. He never needed a leash. He was the funniest unfunny dog I've ever met, and he ended up being a tremendous source of calm for me in the middle of a lot of personal and professional storms.

In the summer of 2003, just a few months after I got Otis, I finally got the call that someone was interested in my résumé tape. It was WEYI Flint/Saginaw/Bay City/Midland. The station covered all those cities but was based in Clio. I drove across the state from Grand Rapids for my first grown-up career interview. What does a natural disaster wear to her first real interview? The highest-waisted pink linen pantsuit on the market, of course. It was summer, so I thought it was appropriate to don the matching Casual Corner low-heeled sandals with the leather carnation between my big toe and second toe. By the time I arrived after the two-and-a-half-hour drive, I learned the lesson that there is nothing about traveling, being nervous, or job interviews that goes with linen.

There's a saying in the Midwest for those raised in a really rural area: "I grew up in a cornfield." Most of the time this is an exaggeration, but as I exited Interstate 75 at Birch Run, I took a left onto a bumpy road that was legitimately in a cornfield. I could see the broadcast tower in the distance, and it was huge! (The tallest in the state of Michigan at the time, I believe.) The street leading to the station off Willard Road was marked by a giant peacock sign (for NBC) that read WEYI 25. I was early, and that was a good thing, because just as I was about to turn at the NBC peacock sign, a train of wild turkeys started crossing the road. Despite the farm surrounding it, it was surreal to pull into a television station where there was an actual possibility that, if I didn't screw it up, I might have a job. The building looked sad and gray, and the receptionist inside looked even sadder, but the news director who had invited me to the interview was glorious.

Her name was Valerie Roberts. Valerie was a tall, stately, beautiful black woman with the kind of short hair that makes you want short hair, even though you know you could never pull it off like she did. Valerie led me back through the hallways, past a real control room and a real studio, and then to the real newsroom. Her office was in the back corner, understated and with one window that overlooked the woods and cornfield.

We chatted in her office and she introduced me to several of my potential coworkers. I left feeling pretty good, and within a week, Valerie called with a job offer. I was going to be Mid-Michigan's newest weekend meteorologist, working the Saturday- and Sunday-night shows as well as reporting three days a week. Valerie told me that she was looking forward to having me at the station and that she would see me as soon as the contracts were signed.

Well, as a twenty-two-year-old with very little professional experience, I hardly read the contract. I saw $22,000 and $23,500 respectively for years one and two and thought, *Sure, why not? I just want to work in TV.* No problem that after taxes that salary would barely cover my mortgage even after I got two roommates. No problem that I would have no extra money to pay for my school loans or food or even a rental at Family Video. No problem that all I would eat on at least five days a week for two years was a bean burrito at Taco Bell for seventy-nine cents. No problem. Because I was finally going to be on television full-time.

I was pretty proud of myself for not screwing up the interview and for getting this dream job. So proud that I let down my guard, and natural-disaster Ginger reared her ugly head and made a huge mistake. For some reason, I thought that since I now had my first grown-up job, I needed a grown-up living situation. As far as natural-disaster Ginger was concerned, grown-ups buy houses when they have grown-up jobs. Never mind that this particular grown-up, me, had student loans and a car payment and an annoying desire to eat every day. And never mind that in my line of work, I was probably going to be moving every two years when my contract came up. I was going to buy a house.

As if my lack of ability to handle a house wasn't ridiculous enough, the house I chose was even more absurd. I thought I'd get a house I could flip and make money on. I knew very little about doing this work myself, but my mom and stepdad are experts, and they would drive across the state of Michigan every weekend for the length of my contract just so the renovation would be done in time for me to sell it and move on. Sometimes I think it's really a miracle I survived myself.

The house was in Flushing, Michigan, and it cost $113,000. It was a corner lot and had a huge backyard for Otis. I put the offer in just days before I had to officially start the job. In my first real-estate lesson, I learned that closing can take up to two months.

But where would I live? WEYI had a solution: they put me up in a bed-and-breakfast in Clio.

So, back in Grand Rapids, I packed up what I believed

I would need for my first week on the job, figuring I would go back to get the rest on my days off and move it all when I finally got my house. I kissed Otis goodbye (I left him with my parents until I closed on the house), assuring him that I would be back for him in just a few short weeks, and made the drive blasting music and feeling like my life was really about to begin.

I arrived late in the evening to the modest house where I'd be staying and was greeted by a nice couple that owned the place. They brought me up to my room and I proceeded to dance in the mirror and sing a few of my favorite songs to celebrate this momentous occasion. In the middle of my "Leader of the Pack" rendition, as I was prepping my outfit for my first day, I realized my shoes were missing.

I ran to my car in a panic, but the shoes were nowhere to be found. All I had were the sneakers in my bag and the flip-flops I had on. Oh, my goodness. I was less than twelve hours from starting my first day of my first salaried job, and I didn't have shoes. It was eleven at night, so all the stores were closed, and I needed to sleep.

So I showed up for my first job in television wearing my favorite pantsuit and flip-flops. The looks I got from my new coworkers varied from snarky to mildly amused to confused. Valerie took me through the station and introduced me to everyone, including the chief meteorologist, Mark Torregrossa, a staple of Mid-Michigan weather coverage, who seemed really happy and grateful to have me on the team—even if I was nearly barefoot. He ran through the graphics system with me and gave me a wink as he coughed and said he wasn't feeling that well. I didn't know what that meant, but I did know I liked him. He had been working for twenty days straight, waiting for them to hire me and was ready for a day off. We talked, laughed, and worked, and all of a sudden I saw the time. It was eight P.M. The day had flown by and the stores were closed again! *Oh, well,* I thought. *Tomorrow I'll make a run to the store as soon as I am done with my second day of training. Everyone already met me in flip-flops. It's not going to hurt to wear them one more day.*

The next morning, I woke up to a missed call from WEYI. Mark had called in sick and they needed me to take his place. It was only my second day, and I was going to fill in for the chief meteorologist for the five, six, and eleven P.M. broadcasts. In flip-flops.

I walked timidly into the studio, and there she was. Erin Looby was sitting at the anchor desk getting ready to do the five P.M. news. She was a gorgeous blonde and was probably wearing a blazer with some fancy label I didn't even know existed. She had perfect hair, perfect makeup, the perfect outfit, and perfect shoes. She sat at the desk with the best posture I had ever seen. She delivered the news in a confident yet pleasant tone, never stumbling on her words. She had been Miss Michigan. You could tell. She turned toward me before that first show started, and not the way most of us turn, but the way a beauty queen's hand turns in a graceful wave. Her entire body turned like that. She gave me a warm smile and said, "Welcome."

Pan to me in my nylon navy suit with the interchangeable collar (pink, since it was my favorite color, of course), far-from-perfect makeup, disheveled hair, and flip-flops. I delivered my first real forecast on a real television station in plastic sandals. Fortunately, the camera stayed mid-calf and above, but still.

The contrast between Erin and me has always been a powerful metaphor for who I am. I have had to learn to love the girl who is just a little off and wears flip-flops. I will never be all together like Erin or any of the other polished women I have worked with in my career. I just need to accept the disastrous part of me, embrace it, find pride in it, and do my best. I am finally getting to that point in my life. And the part of me from the mid-calf up that did the weather that night actually did pretty well for her first live broadcast ever. And it just kept getting better from there. Or so I thought.

Mark returned two nights later, refreshed and grateful. He congratulated me and I felt tremendous. So far, so good. The following week I would start reporting. I sat in the morning editorial meeting, where the staff and management decided what we would cover for the day, and they said, "Ginger, you are going to go to the Buick Open golf tournament and do a live VOSOT for the five P.M. news."

Great! I gathered my things, met the photographer I would

be working with, and then realized there was one big issue: I had no idea what a VOSOT was. I later learned it stands for "voice-over/sound on tape." Basically, it's the part of the news-cast where you see the anchor or a reporter talk on-camera, then talk over video, then stop talking while a complementary sound bite from a person related to the story rolls. But I didn't know any of this at the time, because I hadn't studied televi-sion in college, only meteorology. I felt so stupid. I didn't want anyone to catch me Googling on my computer, so I figured I would just go and see what happened. Goodness, what I would have done for a smartphone back then.

I talked to the photographer who was driving us to the golf tournament to see if I could glean what I'd be doing with-out directly asking.

"So, how do you guys usually get your VOSOTs at the Buick Open?" I asked.

I hoped the mysterious term fit in my sentence and didn't sound *too* ignorant. He explained that he had done the Buick Open for years and thought we should check in with the "weather center" on the grounds and make that part of my hit, because I was a meteorologist and it was going to be very rainy the next two days, which would undoubtedly affect play. Then he told me, "You can have one of them be your SOT."

Outstanding, I thought. *A person must be the SOT! Now I just need the VO. Whatever that is.*

I jumped out of the news van and ran up to the first person

I saw. He was tall, handsome, built, and wearing a black Nike hat. I asked him if he knew where the weather trailer was. He seemed confused and said he didn't know where it was, but wished me a good day. I saw the entrance to the press tent and darted in that direction. By the time my photographer caught up with me, I told him I'd found the directions to the trailer we needed for our SOT.

"That's great, but what did Tiger Woods want?" he asked.

"Who?" I answered.

"Tiger Woods. The guy you were just talking to. One of the greatest golfers today. The guy who won this tournament last year," he said, somewhat incredulously.

Oh. That guy was named Tiger Woods? Not only didn't I care, but I didn't have any idea who Tiger Woods was. In case you're taking notes, being totally unaware of your surroundings and extremely focused on one thing is a calling card for a natural disaster. I like to think Tiger Woods may have found it refreshing that I ignored him. Who knows?

By the time we were done shooting our SOT with the weather folks, my photographer told me he wanted to shoot some VO before we headed back to the station. I was relieved he'd be taking care of the VO. Maybe I was in the clear.

We hopped back in the van. I felt pretty tickled with my success.

It was already three P.M., and we drove back to the station to edit my VOSOT. We were going to be live at five P.M. As he closed the clunky news-van door, he said, "Please write the VO and I will meet you to edit soon." Uh-oh.

As soon as the door shut, I started sweating bullets. Start writing the VO? What did this mean? I pulled out the reporters' notepad that the office manager had sent me out with earlier and started jotting down notes about what I knew about the forecast and what I had learned from the officials at the tournament.

By the time I found my photographer, it was 4:40, and the newscast was less than thirty minutes away. The photographer grabbed my pad of paper and started editing. I am sure he was less than impressed with my bullet points.

But in that nervous moment I learned firsthand that a VO is a voice-over. I watched him match my notes to the video he thought was appropriate.

Eureka! I did it! I had figured it out. I would recite my notes while that video he was cutting played. The SOT would roll separately. It was all making sense.

I went live, and as soon as I was through I ran back to watch the playback. After work I called my mom, thrilled with my successful sleuthing and broadcast. She was so proud of me. I thought to myself, *This is what Barbara Walters must feel like. I love this.*

The next day at the editorial meeting I felt more than ready for my next assignment, because now that I knew what a VOSOT was and had done my first live hit, I was essentially a news expert. But just as I was about to step into the morning meeting, giddy with false confidence, Valerie blocked the door and told me she needed to see me in her office. No matter where you are in life, going to your boss's office will always feel like going to the principal's office, and it's almost never good.

We entered the office, rain pouring on that cornfield outside; my eyes then quickly fixed on another person, an overwhelming woman in the room. Valerie introduced me to Sheree, her assistant news director. As she stood to shake my hand, Sheree towered above me like a giant redwood tree. She was stoic, intimidating, and oddly soft-spoken. I would later find out that she was Valerie's bad cop. *Later*, meaning about one minute later, when Valerie left us together in her office so Sheree could explain that so far, they didn't love my work, and Valerie was disappointed in her latest hire. She was going to take me off the air until they could work with me to get me up to their standards. I was floored. Huh?! I thought I had had a great start. Others had complimented me! I felt just as good as I had doing my broadcasts in Merrillville during college. Again, natural disasters tend to miss things about their surroundings. This was even worse than missing Tiger Woods.

After getting my first real critique, my head was spinning. I felt light-headed. My ears went numb; Sheree shook my shoulder, and it jolted me back to reality. She said, "Let's go to the edit bay and go over your tapes."

In the tiny dark room with two tape decks and two

monitors, Sheree popped in a large Betamax tape air check that held my broadcasts from the week before and proceeded to pick me apart second by second. The old tape-to-tape edit system had the shuttle button, and Sheree hit that thing to pause my tape—every. Other. Second. On everything from my hair, to my makeup, to my clothing, to the way I said "the thumb of Michigan" instead of naming the cities (like Lapeer) in the thumb. It took what felt like three hours to go over my thirty-minute compilation tape, and it was excruciating. Sheree's delivery wasn't exactly smooth or gentle. She made me feel incompetent, ugly, and untalented. Welcome to television. In Sheree's defense, this was the first time I had ever received criticism. If the same were delivered to me today, I would not have reacted as dramatically. I exited that edit bay with the first chink in my armor. Sheree's barbs had been harsh, but I had survived. Barely.

My stomach was in knots, and I forced myself to eat Ritz crackers from the vending machine for lunch before heading back to Valerie's office. She set me to work on her floor with a filing project. That's right, I was not given a chair. It was the cherry on top of the humiliation cake I'd already been eating. I cried the whole way home, wondering if this is what my college Calculus 3 classes had earned me. The next few days were just as brutal, but I sucked it up and took those ladies' harsh criticism. This was the first time I had to take a step back and find strength in myself. I practiced in front of the green screen whenever I could. I would not let them keep me off the air because they didn't believe I was good enough. After all, I was going to be on *The Today Show* in ten years, and neither Sheree nor Valerie was going to get in my way. Ever since I had returned from that internship in Birmingham, I made my password to my Hotmail account my goal in life: *todayshow10*.

I wonder what Sheree and Valerie would have thought of my lofty goal back then. They probably would have laughed hysterically. I wasn't good enough for Clio, Michigan. But I'd get better. A lot better. Maybe they knew that too and it was their way of pushing me. Who knows?

I kept working my butt off at that job and got good enough for Sheree and Valerie to allow me back on TV. Our relationship was strained, but I soon learned I was not the only one they attacked. They went after almost everyone. Even Erin Looby. Which is crazy. Because she was perfect.

Years later, when I got the job at ABC, Erin Looby sent me a delightful congratulatory message. She has always been an incredible supporter and a good friend. She is no longer in the business and is now a mother of four boys, and I am certain she is making the world a better place. She takes the time to write kind notes, telling me when she loves a segment I am doing. She is the real deal. And I guarantee wherever she is right now, her hair still looks perfect.

Through all the drama of those first few weeks, I eventually moved into the run-down house I had purchased, though it needed to be gutted. At least Otis got to join me. Every evening when I arrived home from another battle at work, Otis was my sanctuary. When I would take a good verbal beating from one of the executives, feeding Otis, grooming him, and walking him provided moments of calm within the storm.

One night I let Otis out as soon as I got home at eleven P.M. on a Saturday. He darted, and I heard him bark. Otis never barked. I would joke that he was a mute. So if he was audibly alerting me to something, I knew it had to be very good or very bad. And as soon as I stepped outside and saw his happy face barreling toward me, before he even got to the door, I could smell it. Otis had been skunked. The smell was so pungent. I didn't want him to come inside, so I ran back in the house, leaving Otis stinking on the front porch, and Googled how to get the skunk smell out of your dog. I read peanut butter, ketchup, tomato paste . . . and I proceeded to empty out my condiments on my dog's thick black fur. An hour later I washed and dried him, hoping for a miracle, but he still smelled so bad.

I let him in and made him sleep on an old pile of towels instead of my bed. The next day I received another e-mail beating me up for my far-from-perfect Saturday broadcasts, but it didn't matter quite as much. I was exhausted from staying up playing doggy dry cleaner and wasn't sure what to do. But what happened next marked a real career fulcrum. On my five P.M. show on a Sunday, I was told moments before I went on the air that I would have extra time to fill. You'll remember, we don't have a script in weather, so we can really do what we please, choose our pace, and tell a story. And all of a sudden, right after the current temperatures, it happened. I just burst out with the story about Otis. I talked about the skunk, the peanut butter, and my epic fail as a dog mom. I walked off after my segment was complete feeling like I had let a big weight off my shoulders, but so frightened of my bosses. I had never shared anything personal; I had always done exactly what I was supposed to do. As soon as I got back to the weather desk, the voice-mail light was flashing. And the phone was ringing. I answered. An older gentleman on the other end spoke up:

"Hey, you the lady who has the dog that got skunked? Ha! Well, listen to me. I'm a farmer and I've done this more than a few times," he said.

"Okay," I replied.

"Get rid of all the food stuff and go and buy an industrial-size women's douche. That'll take care of that skunk stench. I promise," the farmer said. "Oh, by the way, I enjoyed your weather report tonight," he added.

I was speechless. I had been beaten down so much that any compliment was as startling as hearing Otis bark. But that was far from the last call I fielded that night. Call after call, e-mails and voice mails, all from people wanting to help. It was a career breakthrough. This was the first time I realized

I am not just a talking head and that I need to allow people into my life a bit. They feel welcome; they connect; they get to share in the simple pleasures or trials in my life, like my dog getting sprayed by a skunk. That Monday morning Valerie called me into her office. I was terrified and thinking, *Well, even though people liked it, she hates me, so she will probably hate it.* She told me, however, it was the best show I had had yet. So relaxed, confident, transparent, and passionate. This is how I should do the weather.

I like to think that Otis was also a bit of a natural disaster, like me. And together, we created a beautiful rainbow.

I will always will be a natural disaster. My first big grown-up job was where I first began to recognize that, although it would be years before I fully accepted it. Nowadays, I believe that you can be a disaster and still be quite beautiful. You just have to find the core of that storm—whether it's being on television in your flip-flops, accepting criticism, or rejoicing in the humor of standing at the store with a giant two-liter douche—and find pride in it. That's where the happiness lies waiting for you.

Chapter Five
WOOD TV

My mother is one of the craziest but most passionate and loving women I know. She has a tendency to walk around this world not only being the best mom, supporting her four kids at every turn, but also being able to start an entire public relations firm focused on promoting her children. Even though I have gotten to the pinnacle of network television and was on *Dancing with the Stars*, she will still tell everyone about my brother and his band, The Outer Vibe, well before she mentions me. They do totally rock, but DWTS, Mom!

She is also a neonatal nurse practitioner who takes care of those tiny babies in the NICU, saving their lives and becoming super close with their parents while the little guys and gals heal and grow over weeks or even months. In 2004, one of those couples happened to be Patti McGettigan and her husband, Bill. Patti was the news director at WOOD TV, the dominant NBC affiliate in our hometown, the station I grew up watching. I was working in Flint, well, Clio, at the time, and my mom proceeded to tell Patti about me. Every day. Patti later admitted she and Bill called my mom "Chatty Dawn," because the woman can *talk*.

And I am grateful she is so vocal, because a year later when my résumé reel ended up in a pile on Patti's desk, she popped it in, liked it, and then realized this young meteorologist from Flint was Chatty Dawn's daughter. She had also heard of me from my boss at the country club I worked at growing up, Cindi Poll—one of my biggest fans in life and a mentor to this day—that I was "one to watch." That combined endorsement made for a great first interview when I arrived at WOOD TV. We had a good laugh, and I felt immediately comfortable. I still had a few months to wrap up my contract at WEYI, but as soon as I could, I sold my house, packed up, and drove back across the state to where it all began.

I was Patti's find, her discovery. I didn't mind that at all, because I knew it meant she had a stake in my success at the station. To say she was the Anna Wintour of Midwest television bosses is probably a bit of an exaggeration, but her directness and complete disinterest in being the kind of advisor who might take you under her wing and show you the ropes was a little intimidating. Still, I liked her a lot, even if I was a little afraid of her. As far as my idols went, there was Brian Sterling, one of the anchors, whom I'd had a crush on since high school, and Rick Albin, the political reporter, who I am pretty sure never liked me. I have a vivid image of that disdain, which manifested in the rapid tapping of his coffee cup every time I pitched an idea in a meeting. Craig James and Bill Steffen were the chief meteorologists, whom I endlessly watched at Dickhead's cottage.

I should have been nervous about starting at WOOD TV, but the truth is I'd been a horse at the gate ready to get going with my career and be the overnight sensation of the on-camera science world since I'd gotten back from that internship with James Spann in Alabama. It only took about one day to wake up from that dream at WOOD TV and realize I was unnecessary. There were five meteorologists at this small station already, and just before I arrived, the one who was supposed to retire suddenly changed his mind. I was, in a word, redundant. But I didn't care. I was starting my career, and that's all that mattered to me.

I imagine a thought bubble above Patti's head when she saw me walk through the door in my Hillary Clinton bright-blue pantsuit with matching sensible heels that said, *What was I thinking?* Had I seen that thought bubble, mine would have read *Game on.* (My thought bubble also probably would have been a bright color and in all capital letters!)

As soon as I arrived, Patti welcomed me and brought me straight to the Weather Center, which is upstairs and separate from the studio and newsroom. It is its own little world where all the meteorologists at the station have their work spaces. A nondescript little cubby full of decades of experience and zest for meteorology. I immediately felt at home. The first person I shook hands with was my idol of all idols at WOOD TV, Terri DeBoer. Bright, blonde, and blue-eyed, Terri epitomizes Western Michigan (we are a very Dutch area). She can be polarizing on television, but the moment you meet this woman, you fall in love. As polished as she appears, her bright pink lipstick often gets on her teeth, and somehow it's completely charming. She is the type of person who forgets her Velcro curler is still on top of her head until moments before she goes on television. Come to think of it, she may just be a bit of a natural disaster herself. That totally makes sense; meteorologists are way too quirky to be perfect.

Terri took me to lunch at the Bull's Head Tavern. I had worked across the street in high school at a city country club called the University Club (think bankers and lawyers, with a place to work out and grab drinks), and certainly never would have guessed that one day I'd be having lunch with the famous Terri DeBoer, my coworker! We sat outside on a beautiful fall afternoon after her noon show. Terri is so recognizable, like a real-life cartoon character sitting on the sidewalk, that we hardly got through the meal. It felt like every other person that walked by stopped, recognized her, talked to her, and wanted a picture with her. And I was with *her*! I was going to be on TV with her. We laughed, and she was naughtier than I had imagined. She said all the things I wanted to say. She was who I wanted to be in the future. I still aspire to be more like Terri.

The next day I was assigned to my cubicle, and it wasn't in the Weather Center, but rather in the newsroom with the main anchors, reporters, and producers. I was surprised to see I was sitting right next to the main evening anchor, Tom Van Howe. Tom was a Western Michigan staple, and to me he was a legend. I introduced myself to Tom, who I grew up watching and idolizing.

"Hi, I'm Ginger. I'm a meteorologist. I mean, I do weather."

It reads better than it sounded. It sounded a lot like Baby in *Dirty Dancing* announcing that she "carried the watermelon." His tan skin, handsome-older-man hair, and glasses made him irresistibly charming and delightful, and he graciously welcomed me.

"We needed a new meterologist," Van Howe said. "That old Bill and Craig are starting to lose it."

I knew he was joking, but appreciated the kindness.

And just like that, I was part of the team at Western Michigan's number-one news leader: WOOD TV.

Every morning I put on one of my three primary-color pantsuits from Casual Corner and channeled all the drive and focus I imagined Diane Sawyer used to fuel herself in the early battles of her career. Every day I pitched story ideas at the nine A.M. staff meeting. My stomach did nervous flip-flops over and over as I stood before people who had been doing this for decades. I consistently got the looks from seasoned reporters and producers, or the finger tapping of Rick Albin. I hated that their annoyance got under my skin and made me question my confidence. I knew that, behind the eye rolling, there was a list of reasons why my ideas seemed ridiculous to these people.

However, I pushed forward. I still made my suggestions day after day, and when I got back to my desk after each meeting, I would make beat calls to local police and fire stations like a cub reporter from an old Cary Grant movie. Around eleven A.M., before the meteorologists and anchors were going on air, I left my cubicle and made the rounds of the newsroom, letting it be known that I was available for any and all assistance (not once had I even been asked to do a coffee run at that point). In the afternoons, I would order the same Jimmy John's sandwich (the club with mustard instead of mayo) and go sit in the station's library to watch old clips of the town where I grew up, the way a medical student might study a cadaver. I saw the evolution of storytelling from long, in-depth, frankly flat storytelling (as if you were reading a newspaper on TV) to what we then knew as the "MTV" brand of quick and flashy editing and graphics. I would furiously scribble notes just to look busy. I finished off the day in my cubicle writing a few imaginary e-mails, as if it had taken me all day to catch up on my correspondence. I was like the mom in the movie *Room*, who kept a very strict schedule to keep her sanity and never lose hope that one day the sun would shine and she would be free. I never imagined that sunshine would be a young, boisterous, six-foot-one-inch-tall reporter named Brad Edwards.

At random points throughout my life, Brad would be my lover, my father figure, my mentor, and my best friend. That's a lot to handle, I know, especially since he is gay (see the lover part). So, let's start with the friendship phase.

About four months after I started, Tom Van Howe suddenly decided to retire, which meant that there was an empty cubicle right next to WOOD TV's newest and least-crucial on-air talent—me. It was a Monday morning. I was making my typical extraneous beat calls, and I felt a force move from behind me, slamming his messenger bag to the floor. I looked up. It was Brad.

"I'm having a cigarette in the parking lot. Let's go," he said.

That's how Brad, my new cubicle mate, introduced himself to me. I wasn't the least bit offended. I'd almost forgotten what a human voice sounded like. I had been a closet smoker in college to stay awake while driving (before I knew I was narcoleptic) and had a social cigarette here and there. So one cigarette with the potential of interacting with a real human within the confines of that WOOD TV edifice on College Avenue was heaven.

Brad is an acquired taste for most—he is blatantly honest with a sense of humor drier than Death Valley. I acquired that taste within the first drag of that cigarette.

Years later, I would ask Brad why it took him so long to talk to me.

"I couldn't decide if your fashion sense was an indicator of your insanity, or sense of humor," he said.

"Bold colors send a message of power and confidence," I replied.

"Oh. So the answer to my question is both."

I wouldn't say Brad took me under his wing, but he did use to say things like "When we get out of here . . ." and "Don't worry, you can trash him in your memoir someday," that buoyed my confidence and made it fun to come to work.

Brad had worked in Lansing, Michigan, before coming home to WOOD TV, too. I say home because he grew up around Grand Rapids. He studied journalism and writing at Michigan State University and had already won Emmys and who knows what else as a twenty-five-year-old reporter. Don't ever ask him to list his awards. Or do. He's won so many Emmys he's started giving them away to his story subjects. Without prejudice, I believe that Brad is the best writer in television news. He has a style and a voice that are unmatched and not seen by nearly enough people. What makes him special, in my opinion, is that his work has changed and touched the lives of so many people. His mission is simple but so hard to pull off: impact. It was inspiring to have a friend who was so respected at work, a friend who would always offer to help make my work better, and a friend who made me laugh until I nearly peed my pants.

Brad and I were like the bad kids in class. It seemed as if we were always getting in trouble for talking too much and for laughing too loud. Rick Albin actually sat in the cube across from us and devoted a lot of energy to throwing us dirty looks all day. To be fair, we were pretty obnoxious. Sorry Rick.

There was the time I walked in and had just had cut my hair short. Like, really short. It was an impulsive move based solely on seeing a network anchor with short hair and thinking that would help get me to the top. It would not, by the way. It made me look round, dated, and just plain funny. But you couldn't tell me that the night I got it cut. In my mind, I looked like a powerful woman ready to take on the world.

I got up that morning for work and styled it as closely as I could to the way the masters at Design 1 (the nicest salon we had in Western Michigan at the time) had done it. It looked a little off and nothing like the professional style, but again, I was perfectly pleased with my new Mariska Hargitay season-five-of-SVU do.

This was a life changer. Note: making large statements

and basing them on absolutely nothing (like a new haircut) is a natural-disaster trait, because a haircut would never be able to live up to my expectations for it.

I marched into the building and flung open the door to the newsroom with confidence. Chin up, I set my purse on my cube and awaited Brad's response. He was looking down, acting like he was actually busy.

I gave the "*mhmm mhmm*" cough that demands attention. And simultaneously, Brad and our friend Brett, who was the morning anchor, looked up at me and froze.

For that brief moment in time, I daydreamed of a world where they were high-fiving me for getting that grown-up do, in awe of my mature beauty. That daydream was suddenly demolished by the piercing sounds of roaring laughter. Brett and Brad were all but rolling on the ground laughing. Brett finally caught his breath and said, "You look just like the little Dutch boy. Why did you take your finger out of the dike? Now the town is going to flood."

More laughter.

I plopped down in my desk and said, "Whatever, it's powerful."

Another shriek and giggle fit ensued. I looked up to see the only thing worse than Brett and Brad pointing and giggling: Rick Albin and the dirtiest look I think he had given me to date.

Brett sat right next to Rick, right across from Brad and me. It was a bad triangle, and I know Rick was over it. But we weren't. We were just getting started.

That obnoxious friendship between Brad and me turned into regular cigarettes in the parking lot and after-hours script help where Brad would mockingly help me understand what a story actually needs, not what the typical reporter sees. Brad taught me to hear, smell, and touch a story, and make the viewer feel these senses as well. He was becoming my journalism mentor.

After a few weeks, Patti started sending me out to report on very small stories like a controlled burn (a wildfire set on purpose) or a rip-current drowning. I worked on the copy for those ninety-second pieces like I was Herman Melville polishing up *Moby Dick*. I still had my sights set on meteorology, but there were zero opportunities upstairs in the Weather Center. Thanks to Brad's support and tutelage, I was okay with that for now. I'd watched enough sports with my dad and stepdad to know that a champion is defined by his challenges, not his victories.

I had no idea, but a defining moment would arrive soon enough with Hurricane Katrina. It was my first big storm to cover, my first gigantic challenge. And it changed everything for me.

Chapter Six
KATRINA

On August 24, 2005, a tropical storm that had been upgraded to a hurricane just before colliding with Florida briefly weakened back to a tropical storm. Then it hit the warm waters of the Gulf of Mexico, using that heat as fuel, rapidly strengthening. It was the eleventh hurricane of that season, and at the time, the forecasts were already looking grim for the Gulf Coast.

As a scientist, I found the storm exciting meteorologically, because we had no idea how tragic it was going to be. So I was elated when my boss, Patti, wanted to send me, along with Dan, one of the station's longtime staff photographers, on a road trip to Gulfport, Mississippi, to cover the storm. This was the break I'd been waiting for. Finally, after six months of biding my time, weathering the low tolerance of my colleagues, trying to fill my days with beat calls and cigarette breaks with Brad, and being the under-the-radar team player, I was getting my chance. Patti was pretty excited about the assignment, too. In her own way.

"Show us what you've got," she said.

Of course I went straight to Brad with this breaking news. He was underwhelmed. I pushed on.

"She has a lot at stake in my being successful. You have no idea the kind of pressure I'm under," I told him.

Brad stood by me as I was packing the van with trail mix, water bottles, and a duffel bag of what I thought was an appropriate "on the scene" uniform—khaki shorts and fitted polo shirts with the station's logo on them.

And then he laughed as I got in the van and started waving goodbye like he was my dad sending me on the bus to summer camp.

"I can't wait to see you blowing down the street in your windbreaker. Watch out, Al Roker!" he cheered.

Dan and I made the drive in a little under seventeen hours. For an older guy who had every right to find me an annoying, overly excited little news puppy, and who definitely had no interest in covering any story outside about a two-mile radius of his house, Dan was pretty chatty. I knew that Dan was a Vietnam vet, but he never talked about those days. His mind had a way of jumping around, like a car with three wheels and no brakes, and I liked him. He accepted me—sort of the way you accept a flu shot every winter, but still.

As the storm made landfall, we were still a few hours from our target: Gulfport. We decided it was safer to wait and go in afterward, so we spent the first night in a motel in northern Mississippi.

As we arrived in Gulfport the next day, all my excitement about covering Katrina was washed away in an instant. It was way hotter and more humid than any place I'd ever been in my life. And it smelled. I still have PTSD about the way it smelled in Gulfport then. As soon as we got out of the van, every single one of my senses was immediately and viscerally assaulted.

The rotting bananas and chickens that had been dumped by the cargo boats emitted a scent beyond description. Because it was so hot and humid there, the body bags that were being collected were beginning to smell of rotting corpses. The refrigerated trucks had not yet arrived. The mold grew rapidly in whatever structures were still standing.

Then the sight—casino barges washed up onshore, as well as strewn-about mattresses, medication, keys, photo albums and books, and wandering dogs. The sounds of mourning and wailing and cries for help from the thousands of victims whose entire lives had been shattered overnight drowned any thoughts in our heads. We didn't yet know that the taste of the precious, warm but clean water that we drank from the bottles we brought would soon be rationed. As the days passed, Dan and I began to limit our drinking to the van—we were self-conscious and wanted to be respectful and discreet, since we still had water to drink. We never took any from the church groups that arrived to help.

And finally, the sense of touch. Actually, this was the one sense that I avoided. There were nails and broken glass everywhere to step around, but the hardest thing for me to do was touch the people there. I saw a lot of journalists hugging the victims, but I couldn't do it. At first I felt I was invading their space. In many cases we were the first people they talked to, the first human contact they had, and here we were, in their faces with a camera and a microphone. I had never done anything like this before, and it felt so foreign, so I was in a bit of a state of shock myself. I don't think I had the wisdom and experience then to comprehend that there was a benefit to our interaction for the victims, too. These were just people, I am just a person, and often the outlet we were able to give them with the camera and microphone was the first bit of therapy available to all. Admitting what had happened. Making it real. And that hug, when genuine, can be incredibly powerful.

In those first two days, I avoided the uncomfortable interactions as often as I could. Instead, I stayed busy studying the science of the hurricane. I measured the waterlines and took notes on the storm's depth, tracking the past, present, and future of Katrina. I tried to occupy myself by doing anything to avoid focusing on the human toll of this disaster, which was just so overwhelming I couldn't begin to wrap my head around it. Dan, who had done this many times, and of course had been in an actual war, pushed me to start talking to people.

We started by knocking on doors, one after the other. Oftentimes, we got no response and moved on. People had either abandoned their homes, died, or maybe were just hiding because of the trauma they'd suffered. When somebody did come to the door, we asked them simply and as kindly as possible if they'd like to talk to us about what happened and how they had survived. I was consistently surprised by the number of people who wanted to talk to us. I particularly remember sitting with one family that had survived Hurricane Camille in 1969. It was unbelievable to me that they were going through this again.

One woman who lived in a house that was heavily damaged invited me to sit down and go through her photo album. It was hard to see the faces in the pictures, because the album was so waterlogged, but I knew they were family members. Her family had survived Katrina, but she brought me over to the shambles of a front porch. All we could see were slabs of concrete where houses once stood and where her neighbors had lived, and where, we would learn, many of them had died.

She carried her album outside with her and wouldn't let go of it, hugging it like a child. Then she looked up from her album into my eyes and started to cry. And in that one moment, that one human exchange with this complete stranger, everything got real. For the first time in my life, I was face-to-face with the impact of a natural disaster. It wasn't just a movie, or even something I was talking about from the safety of a studio. It was impossible in that moment with this woman to keep any kind of the journalistic neutrality I thought was required to be a professional. I simply had no idea what to do. In those very early days of my career, I had no real reporting skills and I knew it. All that brash overconfidence I'd felt back in Michigan, that surety that if I just got my shot, I'd knock it out of the park, suddenly seemed ridiculous. I had no understanding of the balance between listening and guiding people to tell their story. So I just sat with her while she cried. And maybe that was enough. To be a human in that moment, I had to learn to allow tragedy to wash over me, allow myself to feel her experiences and emotions. I was developing compassion and empathy.

On our second day in Gulfport, we went out with the search and rescue team to look for dead bodies. We were using the satellite truck of our Indianapolis affiliate, and they had a hookup with the Indiana task force, a group that had come down with search dogs to look for bodies and make sure all the cars and homes were checked and cleared. Their job was also to go door-to-door and make sure no one still in their homes needed any help.

When you arrive at a pile of rubble that was once a home,

the dog is instructed to go search for bodies or severed parts. When and if they return without finding anything, the task force spray paints a neon orange X on whatever is most visible (a front door, the hood of a car). On each side of the X is a code marking if there are any dangers inside (live wires, gas leaks, etc.).

We saw a lot of those codes being double-checked, which meant that the dead bodies inside had already been removed. The task force was very well organized and helped as much as they could.

After my first story aired that night on the eleven P.M. news back in Grand Rapids, I was criticized online for covering the storm. People had been writing to our station saying that I and the other reporters shouldn't be down there, because we were taking away resources that the people of Mississippi and Louisiana needed. In my tag (the part after the scripted video that rolls), I assured our viewers that we hadn't eaten, hadn't showered, and would never take from those who had already lost so much. Every time I see myself at that moment (when I watch the tape back, even years later), I can tell I was changing. The story could have broken me, but I wouldn't let it. I see such strength in that tag, and it makes me so proud that at that young age, with so little experience, I was not only telling those important stories but addressing the haters in the most mature way available. Since then I've experienced many similar online attacks (people thinking we should stay away from disaster areas), but what they don't understand is that it helps the survivors to tell their stories, or it helps the city or state (when it's a smaller-scale disaster) to make sure a state of emergency is declared so that federal funding becomes available for those who need it. In addition, I also believe it helps inspire people to donate to the victims of such disasters.

Over the next few days, Dan and I filed an average of six short (two minutes max) pieces a day. Katrina taught me how to tell a story. It was my boot camp. If you ask Brad, he'd probably nod and say that's true, and then he'd tell you the story of one of my final Katrina pieces. In my defense, I was emotionally and physically exhausted when I filed the story on a flooded library with a voice-over on top of a book floating in a puddle. Not exactly a submission for the Edward R. Murrow Award, but it was all I had left.

It took too long for anybody to come and help the victims. Everybody knows that now. I saw firsthand that when FEMA finally arrived, they set themselves up under the cover of tents. But it was impossible to hide the helicopters that landed and delivered Outback Steakhouse meals to the agency's personnel. It was horrifying and insulting. We were all starving. Granted, CNN was more prepared for a disaster than Dan and I with our grapes and trail mix, but I know that Anderson Cooper was not eating steaks and creamed spinach on assignment. In fact, he gave me a cup of soup on the fifth day of our coverage. Dan and I considered doing a piece on Outback Steak-gate, but there was no way we'd ever get FEMA to talk to us, and it was too crazy of a story to tell in an impartial manner.

I'd have to say that of everything I witnessed down there (and it was a lot), what broke my heart the most was watching crowds of people standing in line at a CVS that no longer existed. I've never seen people so abandoned, so desperate. For Dan and me, our situation might have been bleak, sleeping in a van and peeing in a corner behind a bank, but we knew we were going to leave.

What is it about humanity that makes people join a line for something that doesn't exist? Is it the lemming phenomenon? Are we just pack animals who know our survival depends on being together? The look in the eyes of the survivors was so vacant—somewhere between a zombie and a character from the HBO show *The Leftovers*. Maybe the dead were luckier. Maybe they were standing in line because a line suggests hope; when you reach the front of the line, you get what you want. And these people had nothing.

Because of Katrina, when Hurricane Sandy hit seven years later, there were water trucks on every corner of Long Island, New York, in ten minutes. Because of Katrina, laws and regulations surrounding natural disasters changed so people wouldn't have to make the choice between leaving their family pets behind and risking their lives to stay with them. Which doesn't mean we handle natural disasters perfectly—we don't. It's just that Katrina was a watershed moment for natural disasters, and it changed me forever.

On our seventh day in Gulfport, a dog dropped dead at my feet. It was a little dachshund, and he was so dehydrated and sick; then suddenly he was just dead. And something inside of me broke. I had nothing left to give or offer as a reporter, and I called Patti and told her I had to come home. CNN gave us enough gas to get to Nashville. When we arrived, Dan and I debated checking in to our hotel for our first shower in over a week, or going out to eat. As soon as we saw the local Outback Steakhouse, we knew the answer.

We ate like starving animals. We drank like sailors on leave. We laughed and talked like old friends who had been through a war together and survived. It was a good, but strange, evening. I knew I'd earned my stripes, and although Dan didn't put it into words, he didn't seem quite as annoyed to be working with me as he had when this assignment began. I had taken the opportunity that was given me, I'd survived unheard-of conditions, and I had done my job to the best of my ability.

When I got back to the hotel room, I had a breakdown that took me by surprise. It was so much worse than the dog dropping dead that had made me come home. I realize now it was a form of survivor's guilt. I got to go home. To a clean, safe place with food and water. Those people didn't. And wouldn't for days, weeks, and even years to come. I managed to get myself into the shower, where it's always easier to cry because nobody can hear you and all the waterworks get washed away. You come out clean. I fell asleep, and we drove back to Michigan the next day.

I was surprised by my homecoming back at the station. Senior reporters I didn't think even knew my name patted me on the back, complimented my work. A few of them even audibly said, "Good job." Even Brad gave me the nod and offered me his own brand of "good job."

"Boy, did you get lucky," he said.

For the first time in my career, I was experiencing the respect of my colleagues. But I felt guilty about it, because it came at the immeasurable price of devastation. Still, as strange as it sounds, I'm grateful I got to witness and cover Hurricane Katrina. I'm grateful I got to meet the survivors and see firsthand the resilience of the human spirit at such a young age. I'm grateful that I got to learn that I am tough under pressure, that I can do any job when it's put in front of me. I have covered many natural disasters since Katrina, and even though I'm a better reporter/journalist/meteorologist than I was back then, I know that Katrina was a baptism by fire, a coming-of-age, a loss of innocence that has made me who I am today.

Katrina also gave me perspective. It made me so grateful for every little thing I had. Everything, down to my shoelaces, meant more to me afterward. Life is so fragile. With that frailty, it's an injustice not to feel happiness and live life to its fullest.

Chapter Seven
BRAD

When I returned from Katrina, my wedding to Joe was less than a year away. But it was at the bottom of my priority list. Joe's sisters and my best friends were constantly reminding me of dates and tasks that needed to be completed. But my mind was so far from the altar. I had seen how fragile life could be and how quickly a real natural disaster could end, or forever alter, our lives. I was inspired to celebrate life and never settle.

Simultaneously, my relationship with Brad was evolving from platonic gossipy colleagues to something more. But remember, Brad was gay. So that "something more" was super confusing.

There were days where Brad would finish his work quickly, and I would have nothing to do. Our usual two smoke breaks became five, and we would play pranks and laugh so loudly in the newsroom that we actually got reprimanded by Patti. She took us into her office and told us that people were complaining and she needed us to settle down or she would have to separate us.

Nothing makes a grown woman feel more like a second grader than that discussion. So, Brad and I obliged, and took our communication to the Internet. We started Gchatting (using a new thing called Gmail) instead of actually chatting. And even though we were sitting right next to each other, something about text versus spoken word made the conversation mature into something more intimate. Our banal chats quickly unfolded into deep philosophical discussions about sexuality, religion, and love. I couldn't wait to get to work every day to engage in our conversations. I would think about our discussions after I went home. I was so inspired by Brad and the dream world we were weaving that I often mentally left the life I was living. In what was essentially the basement of an asbestos-filled television station in Western Michigan, we allowed our imaginations to transport us to a world we created. It was almost as if we were writing our own screenplay and we were the stars. We started expressing our most far-fetched dreams. And then those dreams became hypothetical discussions of our dreams combined. We would banter about what our children would look like, what they would be named. We talked about fleeing the country, quitting our jobs, and moving to Europe. We named our make-believe châ-teau in Italy "La Rondinaia"; we even gave our fantasy nanny, Leonora, a soap-opera backstory. In this illusion, we had driv-ers, chefs, and private jets. This imaginary world was fun, yes, but now I know its main purpose was to fill a major void for us both.

Joe, as I've mentioned, was the absolute best man, sweet, sim-ple, and full of love. Brad, my gay best friend whom I was strangely falling in love with at work, was the most complex, brilliant individual I had ever met. We were falling in love with the idea of being in love. With each other. Even though it made no sense in reality. Or did it? Could it?

Just before Brad moved into the cubicle next to mine, he had lost his father to esophageal cancer. He was a teacher, a sports official, his basketball number is retired and he is in the Grand Rapids Hall of Fame. More than once Brad read me the eulogy he wrote for his father's funeral, and to this day, he speaks about him as the greatest human that ever walked this earth.

When Brad's dad was taken, so was a part of Brad. All he wanted to do from that day forward was to live life like his father, Don "The Animal" Edwards. For Brad, as a gay man, being a husband and father seemed unachievable. But here was this mess of a young woman (me) who may just be interesting enough to marry, have children with, and even pretend to be straight for. Maybe?

I didn't realize it then, but this was an emotional affair. Months passed where nothing physical happened between us. I honestly didn't realize how much impact the emotional affair had on me, because I was never thinking of the life Brad and I had created as a reality. So I kept lying to myself and telling myself that I was making all my life decisions with Joe alone. But after the fateful night/morning when I ran to the post office to remove the invitations from the mail, the real crisis moment occurred.

I had anchored the weekend morning show and was meeting Joe at his parents' house for Sunday dinner. This was the home that their four kids had grown up in, the home where they took Otis and me in as their own as soon as Joe put that ring on my finger. Their family always had Sunday dinner around one P.M. If it were an Instagram post, their Sunday dinner would have read "#FAMILYGOALS."

On one particular Sunday, I came home from work before they returned from church and was up in the bathroom taking off my on-camera makeup when I heard them all come in. I don't remember the exact quotes, but the conversation went something like this:

Joe's oldest sister exclaimed, "Oh, Mom, do I smell glazed carrots? You know I love glazed carrots!"

The other sister: "Mom, you are the best. That turkey looks divine."

And then his dad came in whistling "Zip-A-Dee-Doo-Dah," and not ironically.

And then it struck me. They were too perfect. Not Stepford, freaky perfect, but just perfect saccharine sweet, honey wrapped in stevia wrapped in Sweet'N Low sweet. The problem was that I was not perfect. I was a closet smoking, big-city dreamin', kinda messed up natural disaster, damn it! And there was no way I was ever going to be able to pull this off for a lifetime. I knew it. Now I know I didn't love myself enough to even begin to think I could fit in with their family. Their consistency and joy did not fit the chaos I desired.

All I wanted to do was stand at the top of the stairs and scream, "You people can't be real! BTW, I think I am in love with a gay man and I don't want to marry your son."

Instead, I sat down and ate those glazed carrots. They were really good.

Motivation can come from the strangest source. The next day I drove to work playing my Disney Princess best-of CD, as I typically did (this still happens, by the way, and it's totally not weird). When the first song played, it hit me. I started sobbing and singing aloud to my favorite *Pocahontas* jam, "Just Around the Riverbend."

I feel it there beyond those trees
Or right behind these waterfalls
Can I ignore that sound of distant drumming?
For a handsome, sturdy husband
Who builds handsome sturdy walls
And never dreams that something might be coming?
Just around the river bend . . .
Steady as the beating drum
Should I marry Kocoum?
Is all my dreaming at an end?

That's the tone I walked into WOOD TV with that day. My dreaming was not at an end. I was going to take that chance and find out what was just around the river bend. I was not going to marry Joe.

I sat down and started expressing all of this to Brad online. After a few back and forths about how dull my life was going to be and how I was not Native American, I got up the courage to tell him how I was really feeling. I told him that our château, even if it was in Western Michigan, looked a lot more like the château I wanted to be in for the rest of my life. Instead of wanting to be with Joe, I found myself actually wanting to be with him. I felt comfortable enough to open up and ask him if what we were talking about could be a reality. He said yes. He was feeling it, too. I know now that this was also part of my natural-disaster trait of haphazardness and impulsiveness. I am an expert at jumping before thinking things through.

For a while, being engaged to Joe felt so safe in a "this is right and this will do" sort of way. But I was finally realizing I was not a "this will do" type of girl. The 2.3 kids with the minivan I saw Joe driving in my daydreams had been shattered by the reality that even my gay colleague could make me laugh harder and think more deeply than I ever knew possible. Brad opened a window into a world that I hadn't even known was out there. He challenged me, and that is what I wanted. No matter what disaster I created along the way. This isn't to say I don't think Joe isn't a great and inspiring partner. He just wasn't the partner for me.

So, to tell the story of my calling off the wedding as I did at the beginning of this book without mentioning Brad was slightly unfair. But it would have been super weird to dive in without this background information. You might have put the book down right away. I may have, too. It's a lot to take in. But I promise, as I mentioned, all this chaos of the natural disaster does end beautifully, eventually. So, there, now you know that not only was I finding myself and my inner voice when I called off my wedding, but I was also exploring the idea of a life with a gay man whom I was apparently attempting to date, marry, and buy a château with.

After the engagement was officially over, Brad and I did try to seriously date. Now, as best friends, we laugh about it. And we are both pretty grossed out by it, because he is gay. And he's single, fellas.

Chapter Eight
INTERVIEW IN CHICAGO

My work on Hurricane Katrina definitely bolstered my confidence in reporting and gave me more opportunities to report for WOOD TV. By no means would it be fair to say that I was suddenly crowned Western Michigan's "Queen of All Media"; I still had to fight hard for each piece, but I was grateful for the work.

I had what many would call a pretty enviable job, especially for my age. I would say that my overall state of being was that I was *happy without intention*, until I got the voice mail from Rick DiMaio.

Rick and I had met briefly at a conference in Madison, Wisconsin, about a year before the call. Because he was the chief meteorologist at the Fox station in Chicago and I was just at the birth of my career, our dynamic at that Madison conference was pretty much my being the fangirl to his

Batman. I'm confident that our conversation that weekend didn't go much beyond the merits of the crab cakes, but somehow his *Rashomon* take on that weekend was that I asked him a lot of pointed questions and he was impressed. But his message caught me by surprise. I asked Brad to listen to the message on my work phone so I could watch the look on his face just to make sure he wasn't pranking me. Even then, I had to replay the message fifty times to process Rick's invitation for me to come to Chicago, the third-largest market in the nation, and interview for a job opening as a meteorologist at the Fox station where he worked. My career so far had been PBS to radio to Flint to Grand Rapids. Chicago had been a ridiculous, out-of-reach dream. Until now.

For a girl from the Midwest, Chicago is kind of our nearby Hollywood. I decided I would manage my expectations and just assume it wouldn't work out and that I should be happy just to be invited to interview. It seemed like sound self-talk, but it lasted until the second I saw the Chicago skyline.

The sunlight was glistening off Lake Michigan as I rounded the curve on Lake Shore Drive, and I was instantly transported back to Dickhead's house on Lake Michigan, where my love affair with the weather was born. If I wore hats, I definitely would have pulled the car over and tossed mine in the air as an homage to the ultimate career gal, Mary Tyler Moore. I felt like my life needed to begin here in Chicago, right now, and I needed this job to do it.

I navigated the city traffic in my Mitsubishi Eclipse, sunroof open, like a teenager who'd just gotten her license. Everything about today felt new and I didn't want to screw it up. I parked my car near the fancy Omni hotel on Michigan Avenue, where Fox had paid for my room. My family's idea of traveling when I was a kid usually involved a tent and the great outdoors, so this was my "Eloise" moment. Everything about the hotel smelled like fresh linen and flowers. It seemed as if I were tripping on fresh flowers in the lobby, and in the hallways. When I touched the magnetic key to the door of my hotel room, I exhaled. This is what a hotel experience should be like. This is what *life* should be like. As much as I loved the air mattress I slept on every Memorial Day while camping in my dad's old 1970s tent, I was staring at a king-size bed with pillows the size of ponies and white sheets as crisp as new hundred-dollar bills. The bathroom was even better, with free brand-name soap and a soaking tub that looked like it had come out of *Architectural Digest*. As far as I was concerned, these were the signposts of making it in life. I had arrived.

And then I went for it like a kid in a movie. I dropped my bags and the very grown-up garment bag I had borrowed from my mom and swan-dived into the bed. I had never seen a king-size bed up close before. *Who in the world needs all that room, but who cares, it's all mine tonight.*

Turns out, arriving at the life you were always meant to have is exhausting. After a glorious bath, while wrapped up in my fluffy robe, with a belly full of room service, I hit that bed and its six-hundred-thread-count sheets and thought I'd pass out for the next twelve hours. But then something funny happened, and I was suddenly restless. After a few minutes of luxuriating (a new word that suddenly felt so natural to me), I got up to look out the window at the twinkling lights of Michigan Avenue, and this swelling feeling of light and joy and puppies and Santa Claus started growing inside of me.

Everything is possible, Ginger, it said.

Duh, I replied.

Since I'd arrived in Chicago that afternoon, my life had already been different. Just being in this hotel room, being asked to come interview for this big job, made my old life seem so small. In this moment, staring out the window, I wanted to be walking these streets. I wanted to be coming home from work, picking up Indian food, waving to the guy who ran the local bodega, and looking forward to the weekend Cubs game with my new big-city girlfriends. I wanted to get on with my new life already, because I felt like I belonged in a room that smelled and felt exactly like this. The energy I felt in the city was palpable and kept me awake. This was the magic and fantasy that Brad and I were dreaming about, but now it was a real possibility. Truthfully, I didn't want to live in a château in Italy with Brad and a nanny. I wanted to succeed in my career and live in this great city. There was no turning back. I'd promised myself I wouldn't count on this, I wouldn't want it too bad. But it was too late. I was ready for the next level of my life, and I wanted it to be here.

The only person back at WOOD TV who knew I was going to interview in Chicago was Brad. I talked about it with Joe, but I don't think either of us thought this was going to amount to much. Plus, I was under contract in Grand Rapids for the next two and a half years. It didn't seem possible that a contract could be broken. But obviously it was possible, because I'd just heard the voice say exactly that: *Everything is possible.*

The next morning, I went to the interview at WFLD. Unlike all the other TV stations I had worked at, this building was a beautiful skyscraper right on Michigan Avenue. Rick was waiting for me inside, as well as the news director, several reporters, producers, and a few anchors. Everybody at the station seemed intent on convincing me that they believed I could do this job. I'd never had that kind of experience before. Up until this point in my career, I'd done all the convincing for myself. This big group of successful people took me to lunch at the Ralph Lauren restaurant down the street. I had never been to such a fancy restaurant. You have to take an elevator to the bathrooms, which smell divine. And they don't just have paper to dry your hands; they have the most plush folded, monogrammed paper napkins that feel like cloth, but you throw them away! The place just stinks of elegance and success. Tamron Hall was at the lunch. She was WFLD's morning anchor at the time, and such a ravishing, magnetic woman. She sat next to me and was showing me something on her phone when a text popped up from her friend Dennis Rodman. I felt like I was sitting at the Golden Globes table closest to the stage. Tamron was (and still is) everything I want to be. This shift was happening inside my head, where a life I couldn't have conceived of before was taking shape. It's an interesting thing. I knew the job I wanted; I'd worked hard and set it as a goal for most of my life. But I never imagined talented professionals in my field taking me under their wings, mingling with tangential celebrities, and partaking of expense-account lunches. It was almost too much to process. I realize now that what was happening was simply a full-court press by the station to pull out all the stops to land a young, inexpensive employee. Even if I'd known at the time what a full-court press was, having it happen to me for the first time was like getting a bowl of cake icing for a two-year-old. Pretty awesome.

By the end of the day, when I was riding an elevator up to the top of John Hancock Center for drinks with Rick DiMaio, I was sure I was getting the job. I was literally ascending to the top, thinking of these people I'd spent the day with as my future coworkers and throwing all my managed expectations off the ninety-sixth floor, because I wanted to be in this club. I was the girl who'd just figuratively tried on her first pair of red-soled Louboutins, and I didn't want to give them back. Over my dead body.

Turns out things would get a little more complicated.

The Sunday morning after my interview in Chicago, I reached down to pick up the local newspaper in our driveway back in Grand Rapids and opened it up to the back page (I always read papers and magazines backward for some reason). And there was *my* face, my name, and a headline that made my no-longer-a-secret interview in Chicago look like the biggest betrayal since Brad left Jennifer for Angie. And then my phone rang. It was my boss Patti, and she was on fire.

The picture in the paper was of me walking down Michigan Avenue with Rick DiMaio. My local paper had picked it up from a Chicago paper, where it was news that the station was looking to replace Michelle Lee, WFLD's current morning weather anchor. But back in Grand Rapids, it was news that Ginger Zee was obviously looking to leave small-town WOOD TV for "windier pastures." Clever.

Dealing with Patti's ire was the flip side of the full-court press back in Chicago. No more king-size beds or long boozy lunches with celebrities. I was now facing the wrath of a scorned employer, and it was painful.

I felt guilty, like I'd cheated on Patti. I also felt bad about

Michelle Lee back in Chicago, who, because of me, was finding out she was about to be fired. I met with Patti Monday
morning, and she made it clear that under no circumstances
would she let me out of my contract. I reluctantly called Rick
and the news director at WFLD and explained that I was
incredibly grateful for their hospitality and offer, but I had to
decline. It was heartbreaking. All the fresh flowers, beautiful
paper napkins, and fluffy robes started crashing down as I sat
on the floor of the home Joe and I were renovating. Me, in
a pile of disgusting carpet goo, coming back to my reality. It
wasn't pretty.

I got up off the floor, looked in that cracked mirror, and
said to myself, *This can't be it.* Even though it didn't work out at
Fox, I couldn't shake the feeling that I'd outgrown this Michigan life. I'd seen my new, bigger life in Chicago. I'd felt it, I'd
smelled it, I'd slept with it. It was mine. It belonged to me,
and the strange part about all of it was that I wasn't worried.
I knew it was coming. I intuitively felt that the universe had
used Chicago to show me a bigger part of myself. It was just a
matter of time until it came together.

And I was right. That day, I started getting calls from all
the big stations, including ABC, NBC, and CBS. Everybody
had seen that picture in the Chicago paper, and now they
wanted to know who I was. If WFLD was interested, so were
they. Obviously, they had no idea that Patti had already shut

down my interest in Fox. Within forty-eight hours, I'd gone from thinking I was going to lose my job to fielding calls from every major network in America.

But this wasn't luck. I have always believed that luck is just hard work meeting opportunity. Through all my "lucky breaks," I have worked every holiday, almost every weekend, and more than a month at a time with no days off. Whatever career you choose, whatever dream you have, you just need to be committed, and dedicated, and go for it. I am so lucky to have had parents who, by example, instilled a strong work ethic in me and my siblings since the day we were born. I never questioned that I would be successful if I just worked as hard as they did. My dad worked sixteen- to twenty-hour days, six months a year, enduring difficult physical labor without a complaint. My mom is equally tireless. She is a neonatal nurse practitioner, and she can still save a baby's life after working forty-eight hours straight. She can emotionally care for parents that have lost a newborn and still come home and make her children feel like the most special people in the world. These are the Midwestern, hardworking values that were instilled in me early, and I give my parents all the credit in the world for my stamina, productivity, and work ethic.

When I ended my engagement to Joe, I saw no connection between that and this shift in what I wanted for my career. Intellectually, they didn't have anything to do with each other.

He wasn't trying to stop me from going to Chicago, or any place. But my instincts and intuition were ringing every possible bell and whistle that I couldn't go forward with marrying Joe. Brad had opened up the door to imagining a bigger life, even if it was ridiculous. That one interview and day in Chicago might not have worked out, but they showed me what was possible, and really, that is the biggest gift the universe can give us.

Thanks to my pathological people-pleasing genes, I couldn't allow myself to be too excited, because everybody was mad at me. Patti was still upset about the Chicago trip, and now that she knew I had several other suitors, she was threatening that I'd never be able to get out of my contract. Eventually, her threats softened to something along the lines of "Sure, you can leave, but not without having to pay twenty-five thousand dollars to break the contract," which to me might as well have been asking for a million dollars. And then, of course, there was my fiancé and pretty much everybody in his family, who were all very frustrated with me for my hesitancy about our impending wedding.

I made another trip to Chicago to "visit" WMAQ NBC Chicago. (Stations who don't want to pay for your travel will invite you to "visit if you're in the area.") The visit went well, and I felt confident. But as my personal life blew up with the second and final cancellation of my engagement, I completely forgot about the meeting. It's a weird blessing when all you can do is focus on what's right in front of you, because it creates this space for the dust to settle and fate to step in and sort things out exactly the way they should be.

And that's what happened. The day before I had been supposed to get married, I was in my mom's backyard by the pool with my cousins, kicking rocks and still feeling awful about hurting Joe, when I saw a Chicago area code pop up on my phone. I stepped away to take the call and put on my newsiest professional voice, hoping my midday buzz of Natural Light beer wouldn't do its usual thing of strengthening my Michigan accent.

The voice on the other end of the line was Frank Whittaker, the news director at WMAQ, and not only was he offering me a job plus moving expenses, but he offered to buy out the twenty-five thousand dollars left on my WOOD TV contract.

I hung up the phone and collapsed right there on the lawn. Almost every human emotion was coursing through me, including joy, sadness, guilt, and relief. I lay there for a while and just kept staring up at the sky, thanking God, thanking Joe, thanking Brad, thanking the universe, thanking everything that had happened to lead up to this moment. For the first time in the three weeks since I'd called off the wedding, I felt like everything was going to be okay. I felt good about myself, my life, and my choices. The guilt washed away and was replaced by hope.

It's human nature that even when we find ourselves in a situation where a job or a relationship (or maybe both) just feels wrong, we still hold on. Letting go of what doesn't fit, whether it's people or situations, is one of the most difficult things to do, and it's a skill I am still working to perfect.

The day after the job offer was finalized, I attended my non-wedding with my family and friends, and instead of celebrating a marriage, we celebrated my new job. The spinach artichoke dip was delicious, and the hugs and "I'm so proud of you" speeches from countless friends and family made the day better than I could have ever imagined. If we'd had more than twenty-four hours, we would have refrosted the cake to read CONGRATULATIONS ON LETTING GO AND JUMPING INTO YOUR LIFE.

My mom has this odd knack of allowing the absolute worst photos to make it in to her slide show on the computer in the kitchen that everyone sees. I still see this photo of me on the day of the non-wedding, gaunt, chunky highlights, and a Michelob Light in hand. I'm smiling through my swollen eyes, and in that image of myself I see the beginning of a transition into the woman I am so proud to be today. I also see a woman who thought the drama was behind her, but as the MTV series *Diary* warns, "You think you know me, but you have no idea." Turns out it was only just beginning.

Chapter Nine
CHICAGO

The fanciest store in Grand Rapids was Macy's. We didn't have Whole Foods, and our nicest restaurant was Bonefish Grill. For the longest time, I thought Bonefish Grill was ours and ours alone. I would obviously learn later that it was a chain. We were far from starving, but my stepdad definitely used every coupon in the paper—and still takes great pride in his savings at Meijer (our superstore; it's like Walmart). There really is no other way to define my family and my town except to say we were fantastically middle class. Both my parents worked, our bills got paid, the lights stayed on, and Santa managed to show up at our house each year with at least one thing from my list. My childhood was pretty great, and it was safe, but it was small. This was way before the Internet made the Kardashians' lifestyle an aspiration, the way it is now for so many teenagers. In fact, last year I gave my fourteen-year-old sister a necklace from Tiffany's, and she squealed at the box before she opened it. I never would have known what the robin's-egg blue box meant as a kid. In retrospect, maybe I should have given her just the box for Christmas. (Just kidding, Elaina.) Which is all to say that moving to a big city like Chicago at the age of twenty-five was as exciting as it was slightly terrifying.

I wasn't nervous at all about being able to handle the job. I've loved meteorology for so long that I never question any chance I have to practice it, and I know I'm pretty good at it. But I already knew I was a natural disaster, and I decided to do everything in my power to avoid screwing up, which meant fitting in with the sophisticated denizens of the Windy City I had met that one glorious afternoon when I interviewed for the job I didn't get.

That day in Chicago was all I had to use as a road map for what it was like to live in a big city. Of course it was unrealistic and ridiculous, because when you're being given the full-court press, all the bells and whistles are rung. It's not an ordinary day. And yet, when I thought about how I would blend into life in Chicago, I thought about Tamron Hall, red-soled shoes, linen bathroom napkins, and prosecco-soaked afternoons.

The first thing I decided was that Big City Ginger (BCG) would definitely live in a high-rise apartment. I also decided that BCG makes so much money that she doesn't have time to bother to figure out what her take-home pay after taxes will be, because it's more than three times what she made at the last job, so she couldn't possibly spend it all if she tried. I chose my twenty-seventh floor, Lake Michigan–facing apartment online with about as much consideration as I use to order a latte. Who cares what it costs? What else can I get with it? In the case of my apartment, I clicked yes on the option to have them paint it and make *every single room a different color.* That means on top of my first month of rent plus security deposit ($2,700 each), I added $200 per room, because why not? It looked awful. But I wouldn't admit that to anybody, not even my dad. When he first stepped inside my Skittles Palace, all he could say was, "Wow. You can taste the rainbow in here."

The assistant news director in Chicago strongly suggested that I go see "the best hairstylist in the city." Of course I agreed that BCG needed "the best." I arrived on time for my appointment, already pretty confident with my current layered shoulder-length light-brown hair. I figured this magician would just make me an even better version of myself.

I waited more than thirty minutes before she emerged from behind a black curtain. The room went quiet. Everyone acted as if the "queen of all strands of hair" were holding court. She came to the chair they had prepped me in and didn't even introduce herself. She just started fluffing my hair. Finally, she looked at me in the mirror and said, "It's terrible. Your hair, it's like curtains. Like very bad curtains."

Then she started tousling the top, flipping and flopping it almost violently.

"You need a little more rock and roll. Rock and roll, no?"

I nodded in agreement, having zero idea what that meant and zero intention of asking. She was a scary woman.

She turned me around to work on my hair, and when I was turned back to the mirror to see the final product two hours later, I had to choke back a scream.

She had chopped my hair to about chin length and dyed it raven. I hated it but allowed her to parade me around the salon, boasting to all about how "fresh and rock and roll" I looked. She disappeared at the register, and I was left with a bill of $350. After tipping, it was almost $500 just to get out of there before I had a breakdown. I looked horrendous, like some sort of future sixty-something version of myself. I couldn't believe it.

BCG needed some retail therapy ASAP to make up for this traumatic experience.

I bought expensive down pillows from Bloomingdale's Home, loads of matching accessories for each room (I still have the bright blue creamer from Target that matched my kitchen), and beautiful large red-wine goblets for all the entertaining I would certainly be doing. I also gave myself a theme song.

Every morning on the way to work (a five-minute drive) in the brand-new Mitsubishi Eclipse I didn't need, I sang "Glamorous," Fergie's anthem for single ladies living the life, which of course was me. At the time I bought the car, Mitsubishi had a commercial running featuring four girls singing at the top of their lungs speeding down the highway at night. I knew I'd have my own entourage soon enough, so I'd need a car to take them driving in. In reality, I don't think anybody ever rode in that car with me, and it would be a while before I finally made a friend. But I had my vision of this life I thought I was going to have in Chicago and all the stuff I thought I needed to make it come true. Meanwhile, I spent a lot of weeknights at home puttering around my Skittles Palace making up my own drinking games for *The Bachelor*.

Pretty soon, there was no more stuff to buy, and I was left with something I'd never experienced in my entire life: loneliness. I didn't even have Brad to talk to at night, like *When Harry Met Sally*. He was still at WOOD TV and in the process of moving to Detroit for a much bigger gig, and I'd pulled away and distanced myself, because as much as I cared about him, and as much of a natural disaster as I was, I still knew that a gay boyfriend was probably not a healthy life choice.

During the day, I made building my wardrobe and maxing out my credit cards a full-time hobby. With every new suit from Marciano or wrap dress from BCBG, I convinced myself I was putting another piece of Big City Ginger together. And even though the sales ladies were super friendly and agreed with me that BCG was pretty hot, none of them wanted to be the girlfriends in my Mitsubishi commercial. Nothing in my vision for my new life had hinted at loneliness. I didn't understand what was going on and had no idea where to go from here.

And then one day I got a phone call. A meteorologist I'd worked with at WOOD TV, Bill Steffen, was calling to check in on me and see how I was doing. His thoughtfulness and concern was so parental and friendly and comforting, I wanted to cry. Instead, I told him the truth (that my job was going great) and lied (that I was happy). I think he knew something was up and asked if I'd called his daughter Julie, who also lived in Chicago.

Before I left Grand Rapids, he had suggested I call her and make a date. All this time I'd been in Chicago I'd forgotten to do that. I'm still not sure why. Maybe it's sort of like forgetting there's really good ice cream in the freezer. (Well, I never do that, but some people probably do.) When he asked me to call her, I said I would, and meant it this time. There was no time to waste. I wasn't sure if we'd be best friends, but I knew I needed a girls' night more than I needed that apartment, that car, or even those down pillows.

By the end of the day, Julie had responded to my e-mail, and we made a date to meet for drinks the next night. That part was my idea, and I wondered if I'd been too eager and should have played it cooler and waited for a few days.

Just be cool, Ginger, I told myself. *You're going for drinks, not to the prom.*

Ironically, with all my shopping, finding something to wear was tricky, as I suddenly realized my entire wardrobe was designed to emulate Katie Couric, not Carrie Bradshaw. I probably wore a little too much makeup, because as I checked myself in the mirror by the door to my apartment, I thought I looked like I was going on a first date, not meeting a girl for drinks. Later I would see that Julie had dressed up, too, and both of us laughed about wanting to impress the other.

That night, I walked into the sushi restaurant after valet parking at her condo. I had never eaten raw fish or given somebody else the keys to my car. I was popping my Chicago cherry all over the place and could not have been happier about it.

I spotted Julie sitting at the sushi bar waving me over. She was excited to meet me too! Can you have butterflies from the possibility of making a new girlfriend? Had I just been among the cold natives of the big city for too long to recognize baseline human camaraderie? Who cares! I had girl talk to share, and we were out of the gate as soon as my butt hit that stool. We laughed, ate slippery raw fish, and drank for hours. We figured out we didn't have only her dad in common, but that my oldest childhood friend was also her friend. Sake led to vodka led to wine and back to more vodka. She taught me the only way to eat wasabi—fresh. You have to request it, she instructed. Although Julie was an accountant, she knew all the hometown meteorologists I knew because of her dad. Gossip suddenly felt like air I hadn't been able to breathe since I got to Chicago. There were guys flirting with us, but we ignored them. We didn't want the night to end. We decided to go nearby to the Jefferson Tap & Grill, a bar Julie described as her Cheers.

As we walked around the corner, snow was flying. I mentioned to Julie that tonight the snow was going to be bad, like legitimate-blizzard bad. She rolled her eyes at me, and I realized she had grown up with her dad detailing the distinction between snowstorm and blizzard and was one of the few people on Earth that I didn't need to inform about this difference. (To be an official blizzard, winds of thirty-five miles per hour or higher for at least three straight hours are required. It drives me nuts when it snows a few inches and people say, "It's a blizzard out there.")

I also told her I thought it was weird that I hadn't been asked to cover the blizzard, but chalked it up to being the new gal. Plus, at this point, the beer at The Tap sounded much more enticing than sitting in a truck freezing myself through a newscast anyway.

At The Tap we were the belles of the bar. Her friends were instantly my friends. Drinks started arriving from different tables, being bought for the gals who looked like they were the oldest of friends having the time of their lives. By the time we stumbled outside after shutting down the bar, the snow had piled up to about a foot, the wind was howling, and I knew I was in no shape to drive.

Julie said it was fine to leave my car at her place and asked if I wanted to sleep at her apartment. Flashing my big-city skills, I said I would just take a cab home. I hugged my new friend, waved as I turned, and started walking toward where I thought I lived, desperately searching for a cab. My mood quickly went from jolly drunken happiness, so proud of myself for having a great first friend-date and finally "feelin'" the city, to quickly darting my eyes from side to side at every intersection searching desperately for a taxi. Not only had I not seen one cab in the ten minutes I had walked, but I hadn't even seen another person. It was about three A.M., and the cute boots I had put on with my outfit were soaked through and my feet were losing more feeling by the moment. I had no hat or gloves, and my hair was starting to freeze in sticks. My jovial, inebriated mood disintegrated into fear. I am usually very happy when I have a few drinks, but I can quickly transition into some major drama. That tipping point had been met, and as I started running instead of walking, I was slipping and falling intermittently, crying, and at one point yelling, "Why me?"

Now, this was flip-phone days, not Internet smartphone days, and obviously way before Uber. There was nobody to call besides Julie, and I didn't want to scare her away by revealing myself as such a mess so soon. So here I was, all alone again, under a bridge in the West Loop of Chicago. Because natural disasters come to grand conclusions in an instant, no matter the state of inebriation or weather conditions, I told myself that obviously moving to Chicago had been a bad idea. And then I backed up my conclusions with facts: I was behind in my rent; let's face it, my apartment looked like a bunch of five-year-olds had gotten cracked out on sugar and had a paint party; my fancy new wardrobe would only be fashionable for this season, and I'd be paying it off for ten years; and I couldn't even get to the end of "Glamorous" on my way to work in the car I didn't need. Then I noticed that the L train was as silent as my surroundings. The L train never shuts down, but tonight, due to the snowstorm, it was down. And the only person stupid enough to be outside was a meteorologist. What other proof did I need that I was a complete and total failure? And then I heard a voice. . . .

"Hey—hey . . ."

I turned around, hoping it was an angel cabdriver sent from above. It wasn't. It was a homeless woman chasing after me. And she was yelling. I thought for sure this would be my end. Good. She'd put me out of my misery. I slipped and fell, and when I looked up I saw her.

She was somewhere in her fifties, the lines in her face full of stories I'd never hear, of struggles I'd never know. She handed me the knit purple cap from her head.

"Take it. You need it more than I do," she said.

I couldn't decide what the right thing to do was. I didn't want to offend her, and I was cold. But I also didn't want to take a hat from someone who didn't have a place to sleep. While I debated all these middle-class privileged thoughts in my head, she said, "Fuck it," put the hat back on her head, and walked off. Well played, homeless lady.

I stopped crying and got myself home. And on the way, I knew that things had to change. But I worried I wouldn't be able to follow through on those changes, because let's face it, how many drunk people make big proclamations to themselves that they discard in the morning? I resolved to at least consider the possibility that maybe I could simply relax. Maybe reinventing myself was ruining any chance I had of enjoying this adventure. After all, the happiest I'd been in Chicago had been hanging out with Julie, who seemed to like *me*—the original small-town Ginger.

When I got to my building, I snuck past the doorman, who never acknowledged me anyway. In the hallway outside my apartment, I realized I didn't have my keys. They were with my car keys back at the valet. Of course they were. The perfect ending to the perfect night in teaching me what city life is really like. The only consolation I could come up with was that thank goodness I'd restrained myself from buying a pair of Louboutins, because they would have been trashed.

In the end, I decided that the night wasn't a total disaster. I'd gotten exactly what I'd needed: a raucous girls' night out. I had to admit, I felt better, less alone. But it was time to make some changes.

I moved out of my apartment a few weeks later into something not only more affordable, but actually a whole lot more cozy and me. With a roommate. Chicago was expensive, and the truth was, I was not making that much money. I made a habit of staying away from the mall, but without even noticing it, I realized I didn't miss it at all.

Years later, when I moved to New York City, the same pangs of stepping outside myself to figure out how to create a new version of me came up again, like they had in Chicago. I guess that's normal; there's a reason there are so many great movies about women moving to the big city. The sheer scope of the adventure becomes a very seductive opportunity to start over, to reinvent ourselves. And big cities can be lonely. Those Mitsubishi guys probably understand that better than anybody.

Chapter Ten
THE POLITICIAN

Well before the messy night under the bridge in Chicago, I walked into work for the first day I would be on the air at WMAQ on a Saturday morning in the late fall of 2006. I was hired to do double weekend shifts for this NBC-owned and operated station, covering the six and ten A.M. and five and ten P.M. newscasts every Saturday and Sunday. This was the big leagues. This was what would make my giving up the safety net of Joe and WOOD TV in Grand Rapids worth it. It was an honor to join this newscast, and I was pleased with myself that I felt a little less like a puppy dog than I had at my last two jobs. I was ready to work and contribute to the team. At the time, I completely believed that Ginger Zee, self-proclaimed natural disaster, was ready to be a grown-up. Maybe I forgot that believing a natural disaster isn't always the best idea.

The studio had a side entrance that was separate from the NBC tower but still along the river, and as I entered on that first day, I was reminded of how unglamorous working in television really is. For example, all the on-camera talent were required to do our own makeup, which wasn't a problem for me because I already felt pretty confident in cheekbone contouring and eye-shadow blending. As soon as I walked into the makeup room and met Zoraida Sambolin, the female anchor of the weekend morning show, I knew I had overrated my skills.

I pretended not to stare, but she was a knockout. I set up my little makeup kit next to Zoraida, who was in the middle of putting on her own makeup. This woman's resting face was a gentle, warm smile. All my confidence went out the window, and I fantasized that a MAC employee from the nearby mall would show up instantly to help me pull myself together. It seemed like every TV station where I worked had at least one woman who just shone with the immaculate image I always yearned for but never felt I could achieve. Zoraida was definitely added to the list of ladies I wanted to be when I grew up, along with Erin Looby, and Tamron Hall.

Zoraida introduced herself and peppered our first conversation with the word "girl," like we were old friends catching up on high school gossip. I loved her. We talked about her kids and she asked if I was married. Without filling in any of the dramatic details of how close I'd come, I said I wasn't. I wanted to save that story for when we went out for cocktails sometime in the very near future (I hoped).

"Good, you are young," she replied.

I was no dummy. I heard the subtext, and it went something like . . . *Marriage is not all it's cracked up to be, girl. You keep being you, girl, and wait until it is right, girl.*

When we were done with our makeup, we kept talking as we walked into the studio together. And when I say "studio," I mean like the tiny box of a New York City first apartment, or like a corner cubicle in an office dressed up as a studio. But I didn't care. It was the first day of my new job. I was finally a full-time anchor. No more sitting at a computer sending myself e-mails to look busy. No more pitching stories to all the eye-rolling older reporters. I was going to master that green screen and make all of Chicago fall in love with me. I also made it a goal to take Zoraida's advice and live the single life for as long as I could. The latter goal didn't work out quite as well as the first thanks to another role I would soon take on: environmental reporter.

NBC was big into "Going Green," and as a meteorologist and adorer of the atmosphere it only made sense that I become WMAQ's environmental reporter. I grew up with a compost pile surrounded by acres of fruits and vegetables; plus our first home used geothermal heat. So, it just fit.

One of my first stories, however, brought me into a world where politics and the environment intersect. As my photographer and I headed toward the office of the politician we were supposed to interview, I thought, *This is really it. This is the big city, I am going to ask the questions, get the answers, and hopefully change this world for the better.* So cute.

As we ascended in the elevator, I was not prepared to see what greeted me on the other side of the door. The politician we were there to meet was leaning over his assistant's desk. He looked up with the most gorgeous eyes I had seen in quite some time. He was exquisitely dressed, with a dreamy smile. And as he reached out his hand, wrapping it around mine, it felt like a full-body hug from a long-lost friend—I mean lover. And of course all of my resolutions and convictions about being a grown-up jumped out the window and crashed to their death upon impact. I would have told Zoraida right then and there that she was wrong. Marriage was beautiful and when you know you know.

My head felt fluffy and fluttery and I couldn't concentrate. The only thing I could see was my immediate daydream that had me in pearls and a powder-blue suit walking the steps of the White House with this politician, because he would absolutely be the president of the United States someday. *First Lady Ginger Zee* I saw scribed across the daydream. Chanel suits, children's charities choosing the eco-friendly China for our state dinners. In less than a minute I had it all planned and thought this guy was my soul mate.

That is exactly how volatile I am. That is how quickly I jump—I mean dive—into any and everything. It usually starts as a joke, like, oh, *He is going to be my perfect first husband.* And then I engage . . . aim . . . and fire. *He will be mine.* I had the new awesome career; I deserved a new guy to match, right? Thankfully, my photographer pierced my fantasy bubble with a loud "nice to meet you" reaching out his hand. Then my work brain took over. Thank goodness one of the positive qualities of a natural disaster is that we have different parts of our brains that can work at the same time. So while my reptilian sex/romance brain kept her eye on the politician, my meteorology/career brain did its thing and engaged in what felt like a pretty deep dive into the city of Chicago's environmental issues. I felt good about the interview and even better about my chemistry with the politician. Later that day, my inner dialogue went something like this:

Congratulations! You met your soul mate and have an amazing new role within your relatively new job! Obviously it's a sign; the pieces are all falling into place and your entire life is beginning today! Sex, love, work, apartment with multicolor walls. You are the girl Fergie is talking about—you are GLAMOROUS. If you can just stay focused and show the politician how on point you are, he will get on the love train with you. Just keep it together, Ginger. FOR THE LOVE OF GOD, KEEP IT TOGETHER.

The politician had given me his card and as we drove back to the office I felt it burning a hole in my pocket. I couldn't possibly wait to write him. I figured because it was work related, it was professional to reach out and thank him for the interview so I would have his contact in my "Rolodex" for the future. So, I furiously typed on my Blackberry with the obligatory "It was so great to meet you, thanks for the interview." He didn't wait long to respond; later that afternoon he replied with what I thought was loaded, as far as I was concerned, meaningful subtext. This is basically what the actual text read: "Thanks so much for the interview. I know you'll be great at this job with such passion. Hope to see you at that environmental event or perhaps we can meet up before to discuss."

This is what I heard: *Let's drink, flirt, and probably have sex.*

"Of course, that only makes sense. I'd love to hear more about your policy," I replied.

This is what I meant: *I like drinking, flirting, and having sex too!*

The politician responded quickly again, "Meet me tomorrow night at the Hilton Chicago at six P.M." The Hilton Chicago, okay. I wondered why a hotel bar but went along and said I would see him there.

Arriving early is always crucial when dating in my opinion. That way you are settled and can't have any uncomfortable, sweaty, late entrances. I was almost fifteen minutes early and set up shop, attempting to sit comfortably while looking sexy but not appear as if I had waited too long. He was late. I wrote him a quick text telling him I was here but wanted to freshen up and would be back soon. He said, "No prob, running ten minutes late."

I went in the bathroom at the Hilton, taking note that there was a conference happening as women who had the same "look" as me scurried in and out powdering their noses and applying lip gloss. I looked myself up and down and was satisfied. I was wearing two strands of pearls in an effort to make sure he was seeing the vision I had too. With the pearls I wore what might have been considered a day-to-night look so I could make it look like I hadn't given the outfit too much thought. Rounded toe, chunky heels with a nice boot-cut jean, a gold top that droops in the front, and a brown suit jacket. I was the epitome of early 2000s fashion. I so wish I had a picture to share with you.

When I exited the bathroom, he was waiting for me. *Such a gentleman,* I thought. He grabbed my arm and like a son walks his mother down the aisle, escorted me to the table he had chosen, tucked in the darkest corner. Pretty sure I ordered a scotch, because I had seen the characters on *The West Wing* order Johnnie Walker Blue. They brought the hotel snack mix and we completed the usual verbal dance of *Let's pretend I care about where you went to college* small talk. I honestly can't remember a word he said.

Without knowing anything about Shonda Rimes back then, I was writing a *Scandal*-esque life for myself to live before I even got to the ice cube in my scotch. The politician was just old enough to seem wiser, but not old enough to feel like my friend's dad. So far I was batting a thousand in Chicago for "meetings." This was different from my interview with Fox of course, where I'd seen my career life before me. For one thing, I wasn't picturing anyone at my interview naked. For another, the job interview didn't end with me kissing anyone.

And then I did another thing natural disasters do all the time. Get really mad at ourselves for making stupid choices, especially the ones that seem really great at the time like kissing the politician. Now my inner dialogue went something like this:

Hey! Congratulations! You just got this new role at your new job and you're committing the number-one work no-no! Making out with the subject of your report. You just couldn't keep it together, could you!?

I don't know if that is a real rule, but I knew deep down it wasn't judicious. And it wasn't even a struggle. All it took was two scotch on the rocks and some flattering dusky, back of a hotel bar light, and I was locking lips with a "source" I had known all of two days.

Later that night, back at the Skittles Palace, I decided to give myself a pass. The politician would be my exception. I was still pretty new in the big city, and it wasn't fair how handsome he was. It wouldn't happen again. How about that for a compromise? Instead of telling myself I wouldn't *ever* get involved with someone I met at work, how about I grew up and realized even a glamorous me deserved "one"? But no more. The politician was it and I better stick to it. I poured myself a glass of wine (which I definitely did not need at that point) and laughed.

Of course it won't happen again. There won't be any more men! Chances are, the politican and I are getting married. The search is over! Go GingerPolitican PoliticoSpice?

Whatever. So we didn't have a good team name. It would work; I knew it would.

That following weekend, the politican asked for another date; this time he wanted to meet at a different hotel bar. The invitation came through e-mail again, but I refused to see that as anything but an efficient form of communication. Clearly we were on our way to reading the Sunday *Chicago Tribune* together. Our second date ended the same way the first one had: a brief make-out cut short by him saying he had to work. Unfortunately, it didn't really get much better—we didn't get much better than that going forward.

His invitations to hotel bars ended when he invited me to a real dinner. He made a reservation and told me to meet him at eight P.M. on a Friday night. I had to work Saturday morning, but I told myself one sleepless night was well worth *Air Force One*. It was happening. I knew it was. I could see the headlines: "Whirlwind romance for everyone's favorite politician culminates in European elopement." Terrible headline but you get the point.

Just as I got in the cab the politician texted. He was going to be late. Like an hour late. He told me to head to his apartment in an hour.

I forgave the scheduling snafu because he was probably changing laws and doing very important things. An hour later I showed up at his place. He still wasn't there, but he had his security buzz me up and I waited in his palatial home. It was so refined and not at all like the "bro" apartments of all the broke boys I had dated in the past. This was a real man. The hallway on the way to his living area was lined with pictures of him and other politicians. He had such a special life and I wanted so badly to be a part of it.

He showed up, apologized, and poured me a glass of wine. The past two hours of waiting meant nothing. He was so debonair, alluring, and I felt like he was really listening. I should have remembered that was his job. We had a beautiful night and talked until two A.M., which gave me all of an hour to sleep once I got home. But I didn't care. I was falling so fast. So fast that I promised myself I needed to go slow. I was not going to do anything more than make out. I wanted him to know this was serious. I was serious and a real "lady." I thought as I left, *What would Jackie O do?*

Turns out, over the next few months, I became a whole lot more like Marilyn. I just didn't know it.

I consistently suggested we go out, go to events together, but he said he wasn't ready. He couldn't just jump into things.

I understood that—"the press" would go nuts. We met at his place, had deep conversations (and even deeper Cabernet), and for a few weeks I thought this was so romantic. I would tell myself we needed time to get to know each other; this was what adults did.

But weren't we supposed to go out and have dating montages of holding hands walking through the park, meeting his friends at interesting cocktail parties soon? I started to wonder if he was embarrassed to be with me. My awful "rock-and-roll" haircut had grown out a bit, so that couldn't be it. There was no way to know for sure. But I did know I didn't like this feeling. I told the voice in my head saying this wasn't right to shut up, but it wouldn't. So I decided to separate all the sex/romance stuff and just take this relationship at face value, like the grown-up I knew I could be if I really tried. I thought of when Carrie Bradshaw was messing around with Mikhail Baryshnikov and she kept calling him her "lover," but in a really elongated, exaggerated, and honestly annoying way; and that helped. I decided I was also taking a lover, but not in an annoying way. In a really cool grown-up way.

This could be something new and exciting! A big-city-girl adventure. I let it go on for a few more weeks. We didn't see each other all that often because he was traveling and working a lot and I worked weekends. After a while, I was feeling icky whenever I saw his name pop up on my phone. Not just like, *What is wrong with me that he won't be seen in public with me?* and

What's the matter with me, why aren't I worth dinner and a movie?
but I even started thinking he was doing this to lots of women
all over the place. At least that's what I told myself he was
doing. I never asked. That would be too mature. I was young,
naive, vulnerable, needy, and not ready for any relationship
after the end-of-my-engagement-to-Joe and dating-my-gay-best-
friend ordeals.

You'd think that all of this would have been enough for
me to end it with the politician, but I took our relationship
on as some kind of challenge. A challenge to be special, to
climb the ranks and step over the other girls he'd done this
to, like the big screen at a Flywheel spin class where everybody
is pedaling like lunatics to beat each other on a stationary
bike that goes nowhere. Because I was the type of girl you
want your friends to meet, I was the girl you definitely take
home to Mom, and I had always been the girl that you want
to girlfriend-up as soon as possible. At least, that had been my
experience until this point in my life.

But, before my inner voice could scream loud enough for
my real voice to say something, the "race" ended. In Chicago
there are a few magazines that highlight the social calendars of
Chicago's "celebrities." I had had a few photos at events in the
magazines and always found it fun to flip through the glossy
gala and charity ball images. As I was getting my nails done
with Julie one day, I flipped the page and my heart sank.

There he was. In a photo with a stunning woman. The caption read, "Politician and fiancé xxx."

Suddenly, my daydream returned. This time it had me slapping him across the face in the oval office, tearing my pearls until they burst all over his beautiful wooden desk and storming out to go run for office myself. So much like Mellie Grant on *Scandal*, whom I would compare myself to years later.

I filled Julie in on the secret relationship I had been having because I no longer needed to protect him. *What a dog*, I thought. Not only was I really being hidden, but I was the side chick. I had never been that before. I was indeed Marilyn Monroe to his Jackie O. But that wasn't who I saw in the mirror.

It's amazing how long it takes us to do the right thing for ourselves once we know what that is. I wish I would have been strong enough and loved myself enough to end it when I felt like I wasn't being treated as I should. Instead it took a photograph in a magazine to make me do the right thing. Ultimately, my relationship with the politician didn't end dramatically. It didn't even end like a spin class does, where you just get off your bike, grab your clothes, and go home. It just sort of faded away after I saw that picture. Honestly, I wish my other relationships had ended so smoothly. I wasn't even mad at him. I was more mad at myself for asking so little and expecting so much. The truth is, the politician had never led me on about what he was offering, he had just hidden the fact that he was engaged to be married. I made all the other prom-ises and expectations for him. And I didn't do my research. I was still coming to terms with what being a grown-up meant. I thought it meant I'd be cool just casually dating, but it was starting to mean something much more complicated. I had to start discriminating between all those voices that were telling me a complete stranger was my soul mate, that having just a physical relationship was okay even when it made me feel like crap. I had to start listening to that voice that was telling me it didn't feel good. Because that was my grown-up voice.

There is no way I can skip over the part about how this made me feel after it ended, even if I wasn't mad at him. I did feel rejected and depression was always looming. To make myself feel better, I often went full force back into the dating pool, allowing any attention that would come my way to enter. Instead of working on myself, I got into a terrible habit of pretending that these instances meant nothing to me because I would just move on so quickly to whatever I could get my hands on next. I had no love or belief in myself, I needed men to give me worth. That is a difficult lesson to learn, and it took me years. I am not one of those people who thinks you *must* spend time alone, because I think people are meant to be together. But I do think you need to respect yourself enough to find the right person or just be cautious before diving in because there is no value in allowing others to determine your worth. Even if they could have elevated you to First Lady-fashion icon status.

Chapter Eleven
TROLLS AND TV

Once I was out of my fancy apartment and had moved in with a roommate to save money, settling in to WMAQ in Chicago was easy. I quickly got over the first-time-on-air-here jitters and started getting into my groove. I worked doubles on weekends, every holiday, and often for twenty days straight, as the other meteorologists at the station had loads of tenure and vacation time. I was always happy to fill in, as it meant more exposure for me. On the flip side of that good news, more exposure on air meant more face time with a demographic I like to call the "MV," or *mean viewer*. I'd met them before in Flint and Grand Rapids, but there I'd been able to count all of them on one hand. Suddenly, in the big city, they appeared to be multiplying like jackrabbits. Actually, they were more like rabid gophers than fluffy rabbits; these MVs had sharp teeth and wanted nothing less than to tear out a piece of my soul. I'd call them dickheads, but we know there's only one true Dickhead.

In my early days at WEYI and WOOD TV, there had been the occasional call from an MV who felt compelled to share their distaste for my outfit or correct an egregious mispronunciation or grammatical mistake. The barbs of my small-town viewers felt like cotton balls compared to the *Game of Thrones* weapons the Chicago viewers were tossing at me on a daily basis.

Here's an example of an e-mail message that arrived from an MV just after I started my job.

> We find Ginger Zee to be hard to follow, inconsistent, confusing, incompetent, and generally annoying. Has anyone at the station ever even LISTENED to her??? Our family finds her so awful that we are switching to Channel 2 CBS for our local news. Unfortunately for you, a number of friends and other family members feel the same way that we do. Glamorous, sure. Intelligently competent—HELL NO!

Even worse than this woman's message was that she sent it to the general mailbox at WMAQ, which meant everyone I worked with could read it, including my colleagues, the interns, and my bosses. The first time I saw the e-mail, I did exactly what they tell you not to do in Women in Business 101—I cried. I stammered, bawled, hid in a closet, considered a career as an underground tollbooth collector. It was awful.

Social media then (mid-2000s) was present, but nothing

like it is today. Today, all the MV attacks, from whatever the source—Twitter, e-mail, Facebook, Instagram—arrive instantly. At least back then, you had to open your work e-mail to get attacked. I am so grateful that my early days in Chicago were before Internet trolling became a national pastime and that I had time to develop a thicker skin.

I waited to hear from my coworkers about the e-mail. But it was weird, because no one said a word. I started to get a little paranoid. Were they all talking about the MV comment but afraid to talk to me about it? I finally got up the courage to talk to Zoraida. She said she hadn't heard about the e-mail, then laughed and told me to get used to it.

Great. What I'm sure she meant to be comforting was now going to keep me up at night. She went on to explain that I basically wasn't in Kansas anymore—this was the big city, the big leagues, and I read into this that I needed to be a big girl. That made sense, so I resolved then and there that I would have a thicker skin than a rhinoceros. I would be so tough, the senior anchors would come to me for advice on nasty e-mails. I love my job and I'm grateful for it, but it's hard being the punching bag for Internet trolls, who mostly just want to take out their own frustrations on a stranger who they probably think makes too much money and has a really easy job being on television every day. Over time, I have developed a strategy that combines kindness with an underlying hint of sarcasm and the kind of fierceness I imagine Beyoncé's alter ego, Sasha

Fierce, would want me to employ. Here's a recent example of an MV and my response:

> @ginger_zee if you wanted a life of travel, why did you have a child . . . does your child know who u are or is it all about u.

That's it. A firm and resolute condemnation of my parenting based on the need to travel for my job. Good times, right? My husband wonders why I engage with these cyber thugs, but I can't help it. I take it as a challenge to meet their vitriol with kindness. It's important to me to remind them that there is a human at the other end of their keyboard.

So here was my response:

> Proud to be a mom who can do both. It's amazing what we are capable of.

Because of the 140-character limit, I sent a second Tweet that read:

> Also, my mom worked a ton and we are now best friends. Have a better day!

Typically, my "aim high" strategy shuts down the conversation, or they suddenly turn into my greatest fan.

It makes me sad that almost every single MV I get attacked by is a woman. It's a cliché I've heard for years that women are meaner to each other than men are to us, but I didn't want to believe it. Unfortunately, as a scientist, I now have loads of empirical evidence in the form of e-mails and social media comments that back up this horrible cliché. These comments are never about my brain, but are always about my outfits, my voice, or my apparent raging on-air sluttiness. Here's another example from Chicago:

> Where is Andy Avalos [popular WMAQ weatherman]? And who
> sprays on Ginger Zee's clothes?

For a recovering anorexic, this was a tough one to read. Because my sick brain immediately saw this as *Ginger Zee looks fat.* I had to fight my inclination to write back, *I had to buy these clothes myself, lady. With my own money. And TV adds ten pounds. And I can't imagine what you are wearing and look like.* Instead, I wrote:

> Happy Wednesday! Thanks so much for taking the time to write
> in and share your opinion. I am guessing you mean my clothing
> appears too tight? If so, I am sorry it came off that way and by no
> means do I intend to offend. I am just a young woman who studied
> meteorology only, no TV, trying to figure out fashion and flattering

cuts like the rest of the world. If you happen to have any specific help or suggestions on where I might find affordable clothing that you think would look better, I am all ears. I am relatively new to the city and hope to hear back from you soon. Above all, thanks for watching and I hope you have an outstanding week.

Postscript: She wrote back with several very specific sartorial suggestions, including that Lord & Taylor has their semiannual sale every March and September, and that Talbots makes elegant, loose-fitting pantsuits—and that they have a 40 percent coupon. She also became one of my most loyal fans and supporters.

I try to think of these MVs as sad or angry folks at the other end of a computer who think I have a closet full of amazing shoes and the easiest job in the world. And I get it. My job comes with the perception of glamour, fame, and fortune. It's not exactly true, but I do understand the attitude. I'm lucky. But does that mean I have to become a target of anybody with a laptop? I've never strongly empathized with the Gwyneth Paltrow it's-so-hard-to-be-a-mom-and-a-movie-star rant, and I guess I accept that MVs are just part of the job one must endure. So I do the best I can to handle each comment with as much empathy and this-isn't-about-me kind of attitude as I can.

Very rarely do I get an e-mail or post about something I

actually do for a living—like forecast the weather. This one got close:

> I watch NBC 5 all the time,,, love the younger generation,,, but it is very annoying to me when Ginger Zee, falls into that–gravely voice thing,,, I find it very unprofessional, at times it seems very forced. The audience is more then the 20 year olds.

Here is my response:

> Please help me understand what you mean by "gravely voice thing." I take all comments very seriously and always want to come off as professional. Did I sound like I was in a grave, and if so, any suggestions on how to remedy that would be most appreciated, as I am very much alive.
>
> I appreciate you taking the time to write and for watching,
> Ginger Zee

Here is her response:

> Hi Ginger,,, the only way i can explain it is,,,, a deep throat thing,, and it sound very gravely,,, To me it is an early 20ies girl thing,,, they force their voice to go low and it comes out "like" gravely. Thank you for getting back to me,,, you are such

a talented and attractive young woman,, I wish you well in your career. I love channel 5 and if you have any pull,,,, I wish they wouldn't show so many programs in the wide screen,,, I love to see the top of the heads of people (or should I say, the entire head)

See how she turned there? Started complimenting me? Always fascinating.

Then there's the straight crazy:

Sent: Saturday, February 13, 2010 5:30 PM
To: Zee, Ginger R (NBC Universal)

Subject: Honey are we retaining water tonight at the auto show or did you eat some bad Blow Fish to make that Punkin head of your look fat on TV tonight ?

My response:

Pumpkin head? That is a new one. Thanks for the chuckle.

Thanks so much for watching!
GZ

No matter what you do, you always end with "thanks for watching," because that is the point. They were watching.

They have a choice. They chose to write after disliking what they were seeing, and then chose to keep watching; and I will always be grateful for that.

While I was pregnant, I got this doozy of an e-mail:

Subject: Dress with respect

You dress like a sausage. How embarrassing. I'm sure you get paid plenty and can afford to buy clothes. Disgrsce

And yes, she spelled *disgrace* wrong. Misspelled words and poor grammar are almost a guarantee when an MV writes. Very rarely, I would say 0.01 percent of the time, do they have proper grammar and spelling.

Here is my response to the pregnant-woman basher:

Oh no! I am so sorry you think I look like a sausage. That is definitely not my intent. As a very pregnant woman, I am doing everything in my power to not look huge. I guess I need to work a little harder.

Thanks so much for watching and have a spectacular day,
Ginger Zee

This may seem strange, but I get excited these days when I hear from an MV. It's weird and maybe slightly masochistic, but it reminds me just how far this natural disaster has come. I'm not saying the attacks *never* hurt, but mostly they are amusing. I still respond to every one that is appropriate (meaning, I can't picture them in a dark room lit only by the screen of their laptop, framed by clippings of me taped to the wall in the background). A therapist once taught me a terrific tool to use in handling the attackers, which is to separate others' feelings and opinions from my own. Everyone is entitled to their feelings, and my only job is to acknowledge (not change) their feelings using the template, *I'm sorry you feel that way, but I love . . . matching my shoes to my dress* (for example).

I suggest this tool for anybody facing any kind of harassment, undue criticism, or bullying. For some reason, honoring the feelings of your attacker disarms them, and they either go away or come back with such a lame response that it's game over.

I have also started utilizing my "Good Guy Team" (also known as my fans; let's not forget, they exist, and I love them all) to bully the bullies. For example, I got a simple *Your the ugliest weather girl I've ever seen* (yes, *your* was used improperly) post on Facebook a while back. I quoted it (without blocking the person's name) and informed them, *Please get it right; I'm the ugliest meteorologist you've ever seen.* And then my fans took care of that guy. I believe he had to shut down his Facebook page that very same day. Not that I intended for that to happen, but it's important to remember that every single one of us has our fans, our friends and family, and we need to let them step in and do the heavy lifting once in a while.

In the case of the your-the-ugliest message, I'm fine with their opinion that I'm ugly. Call me a weather girl? Oh, hell, no. I have studied too much, worked too hard, given up too many days off, and made too many sacrifices (happily) throughout my adult life to be reduced to a stereotype that diminishes my talents and skills. I believe that each of us has a responsibility to take seriously the power of our words. Otherwise they become dangerous in the hands of those whose intent is to manipulate language for their own (evil) ends.

While my responses haven't changed too much, my reaction to these e-mails has. Thinking about the initial mean viewer in Chicago, my tears, and running to Zoraida, I realize now that all I was looking for was validation. That I was okay. But, thankfully, growing up has taught me that it's my job to validate myself.

Here's the bottom line: you either believe them all, or don't believe any of them. Because, to be fair, I get plenty of complimentary e-mails, tweets, Facebook posts, and Instagram comments as well. People blow up my skirt all the time. But I have chosen to not take any of them too seriously. Not the good or the bad. You can't. All you can believe in is yourself.

One of the ways I dealt with the trauma of that first MV

attacker back in Chicago was starting a folder labeled BOOK WORTHY in which I put every single MV comment I've ever received. Occasionally I find myself wandering through that folder and am amazed that time really does heal all wounds. Some of them are hilarious. And even the ones that aren't, that are just plain nasty, don't sting anymore. My shrink was right—everybody really is entitled to their opinion, and I don't have to take it on; I don't have to internalize it. Turns out being a grown-up does have some benefits.

Now, I'm gonna get my fat pumpkin face some sprayed-on clothes and see you all on network television.

Chapter Twelve
TEACHING AND SCIENCE

Almost a year had passed at WMAQ, and Chicago was really starting to feel like home. It was the summer of 2007 when I got a message from my old friend and former classmate from Valparaiso University, Paul Oren. Paul and I had been resident assistants years earlier at Valpo, and he was now teaching at our alma mater.

In his e-mail, Paul was telling me that Valpo had a class called Weathercasting that they had recently developed, and he thought I would be the perfect professor.

"Ging, this is an adjunct position. One day a week. You would be brilliant."

"Stop. What is adjunct?"

"Don't worry about it. It means you don't have to have a master's."

"Okay, good. I don't."

And by the end of a short and sweet conversation with the same guy I used to patrol the floors with at Alumni Hall, I was taking on another job: adjunct professor.

I walked into that classroom in the fall of 2007 with such determination. I was wearing one of my red Ann Taylor power skirt suits with fresh-out-of-the-box matching kitten heels. I was so proud to be returning to Valpo just five years after graduating, especially because this time I was returning as a teacher. The doors of Schnabel Hall swung open and the smell of books and labs and maps washed over me like a warm breeze. Even though I had studied meteorology in a different building across campus, this one had a special memory for me.

Less than a decade ago, I had been a campus radio cohost broadcasting from this building. It was initially my friend James's show, and he had asked me to cohost. The timing was perfect. I had recently decided that I was interested in being a television meteorologist. I was working on my science degree, but I didn't have any education in, much less practical experience with, being on air. The concept (and I use that word loosely) for the show was a mix of random music and talk-show-lite ramblings by me and James. Our audience probably never broke double digits, but we didn't care.

The small audience allowed us to grow more comfortable speaking into the microphones. But after a few months, we grew restless and began to wonder what we could do to increase our listeners. After a blue-sky pitch session (a term that means nothing is off the table and no ideas are dismissed out of hand), we landed on what was probably our craziest idea. We were going to rebrand our broadcast with the provocative title of *Topless Radio*. We hung newspaper over the glass booth that other students passing by could see through to give the illusion that we were both buck naked from the pants up inside the studio. We weren't, of course, and I seriously still can't believe anybody believed us, but they did. Very quickly our show became a huge scandal at our Lutheran university, and as a result, our audience grew. Somehow we managed to stay on the air as *Topless Radio* for the entire spring semester without the administration shutting us down. Now that we had at least a few dozen listeners, James and I felt like legitimate radio personalities. The campus paper wrote a story about us, and we were asked to participate in a local Jell-O wrestling event for charity. And it all began in Schnabel Hall, where I was about to deliver a lecture on "weathercasting" to a roomful of mini-mes.

These kids were lucky. When I was in school, we had to find our on-air experience for ourselves out in the real world, which is pretty impossible. I was jealous of the opportunity they had to take this class, and I was looking forward to playing a role as teacher and maybe mentor.

Leaning against a desk next to my lectern was a copy of my very first billboard as a professional meteorologist, wearing an electric-blue blazer with black buttons. Even in that terrible pantsuit, that billboard made me feel like a prodigal superstar. It didn't matter at all that there were only six students in the class. I was going to pour myself into each and every one of them and make them all get jobs. (Postscript—I am proud to report that today four of the six students from my first class are currently working as television meteorologists, and many more from the six semesters I taught are in television.)

Only one of the students was female, and I knew the moment I met her that she was going to be a rock star. I knew she was going to fight the same battles I've had to fight just because of my gender. Her name is Ellen Bacca, and she would later become my intern and prodigy. I used to joke that she had so much talent that I was going to teach her *almost* everything I knew so she wouldn't take my job. Ironically, today she has my old job at WOOD TV in Grand Rapids and is teaching weathercasting at Valparaiso.

When I was a student at Valpo, the emphasis was on going on to grad school to get your PhD. That's it. The path was set for a lifetime of academia, or working in-house for the government or a big corporation. I don't believe anyone from my graduating class at Valpo went into broadcasting but me. And that's fine. I've always been in the minority as a woman interested in science, so I feel compelled to speak for and support that minority. And even though it's been a few years—okay, a few minutes—since anybody's called me a "weather girl," I know it's still a default title for any female meteorologist, so I fight the good fight. I used to wonder why nobody ever says "weather boy," but lately I've started to think of it as a non-gender-specific pejorative. Nobody who loves science as much as I do, who spent four years studying differential equations and linear algebra, deserves to be reduced to a sexualized cliché.

Unfortunately, I have had to fight the prejudices many harbor toward television meteorologists since long before I even was one, and it began right here at Valpo. One of my professors, whom I admired and respected, told me that a career in television would be nothing more than a "waste of my brain." It was an awful feeling at such a young age to be told this dream I had that filled me with excitement and purpose was basically stupid. Because, for whatever personal reasons he had against television, he refused to write me a letter of recommendation for a scholarship I was applying for.

"I used to be on the board of this scholarship. I know what they are looking for. And it's not you," he told me.

I asked what it was that they were looking for.

"Someone more academic."

Seriously? So on the one hand, he thought I was too smart to be on television, but on the other, I was too dumb to be seriously considered for a science scholarship. Welcome to the paradox of being a smart woman.

I wanted to tell him *he* was too stupid to be molding the minds of the next generation of scientists, but instead I simply took my application a few doors down to a female professor who was happy to write the letter of recommendation for me. It makes me sad to think that, at such a young age, I had already accepted that men would think it was okay to dismiss me so easily. But at least I was smart enough to figure out how to go to the right person to get what I wanted. As a punch line, a week later I brought him an application for a scholarship from Glamour magazine, just to see if he would write that. Of course he was happy to.

Which is how that one young woman in my class, Ellen Bacca, fueled my passion to teach. To this day, she consults me before she makes any big career moves, and although the months of wisdom I imparted to her in Weathercasting and Weathercasting 2 were, I believe, helpful, I think it's our relationship and my commitment to being her mentor that matters the most to both of us. Just last summer, she and her husband came to my parents' pool while we were home visiting. I felt such pride talking to her about her career and how much she felt I was a part of her journey.

I made sure every student I taught, especially the women, knew that we are no less scientists just because we choose to express science on television. Yes, I wear brightly colored dresses, high heels, and lipstick, but I'm also a Certified Broadcast Meteorologist by the American Meteorological Society, a distinction that is held actively by fewer than 350 television meteorologists in the country. (The certification process requires having a BS in meteorology, passing a difficult test, and submitting a tape that must be approved by their committee.)

People still think it is okay to call me a "weather girl." It's not. But I get it. When television news was just starting, weather girls were women who recited the information created by meteorologists at the National Weather Service. We've come a long way. Today, almost everyone you see doing weather has a science degree and is genuinely interested in the atmosphere. The job is just too hard and too challenging to pursue if you don't love the weather.

I am the first female chief meteorologist at a network. And I want to scream it from the mountaintop, or at least Times Square. The crusade to be taken seriously as a scientist despite my gender and to celebrate all women who love science as much as I do is one I am committed to. I have hope that one day the term "weather girl" will be as obsolete as "housewife."

I was determined that this was not going to be a class where I droned on about theory, reading, or problem sets that my students would never use in their careers. More than anything, I wanted to give them practical tools that would make their transition from college to career a little easier than mine had been. To that end, I made them come to class in on-air clothes, trading in their comfy college sweats for suits, ties, and pantsuits. I taught them how to apply for a job, how to negotiate a contract, and how to gracefully exit a job when the time was right. One of my fondest memories came in my third semester of teaching, when I had eleven guys and Ellen in Weathercasting 2. I made them all sit around the lecture table in front of Mary Kay makeup tutorials. I will never forget the image of eleven guys trying on makeup.

I loved watching the students get on the green screen for the first time, which started the very first day of class. What I found was that some people are just inherently talented, but there are aspects of being on air that can be taught and practiced. Here are a few examples of TV oddities that folks in news commit.

Vagina Hands

Vagina hands is that odd way of holding your hands when you are on television. I don't know where it started or why it hasn't been abolished, but in my class, it was. I would always tell them, *Stand like you do at a party, like you do naturally. If you wouldn't talk to your mom holding your hands in a diamond shape like a vagina, then don't do it on TV.*

Crutches

The crutch is a reference to something a person does or says when they get nervous on television. It's the "Uh . . . um . . ." or "As you can see . . ."

I was ruthless with my students when it came to ridding them of these habits before they got too ingrained. It was inspiring to me to see them build their confidence without their crutches. Everybody responds to confidence. It's exciting, it makes us trust you, and it's a huge component of being successful on television.

Overused Words

There are many of these, but a few of my pet peeves are *residents*, *motorists*, and *massive*. Motorists? Residents? Aren't they just people? The word *massive* has been so overused, it has lost all meaning to me. Massive sinkhole, massive manhunt, massive storm . . . so what really *is* massive?

To this day, I get calls, texts, and e-mails from former students asking for advice on contracts; or they want to share the great news that they got their dream job. My former student Sean Bailey just got his third job, the one he has been gunning for his whole career, at the ripe old age of twenty-nine in Myrtle Beach, South Carolina. I love hearing from my former students like Sean because it always reminds me that I have a purpose to pass on what I have learned, and that I can make a difference in people's lives. Everybody loves that feeling.

On his way to his first day at that dream job, Sean texted me:

WHEN YOU STEPPED INTO *GMA* FOR THE FIRST DAY OF YOUR DREAM JOB, ANY ADVICE FOR ME AS I STEP FOOT INTO MY DREAM JOB?

I responded:

THE MOST IMPORTANT THING IS TO REMEMBER THAT IN A MONTH IT WILL BE LIKE
EVERY OTHER PLACE YOU'VE EVER BEEN. NOT TO TAKE AWAY FROM THE MAGIC,
BUT TO MAKE YOU LESS NERVOUS. YOU BELONG THERE. YOU ARE GOING TO BE
AWESOME.

He sent me this note at the end of the day:

THANK YOU. TODAY WAS OVERWHELMING, BUT IN A GOOD WAY. ALSO, I WORK
A SPLIT SHIFT TOMORROW AND GET TO GO TO THE BEACH IN BETWEEN. . . . HOW
COOL IS THAT?!

That is cool, Sean. And that is what I miss about teaching. Sean still thinks working fourteen hours with a four-hour
break in between is "cool."

Teaching allowed me to remember the purity of this business, my passion for meteorology. Hopefully I'll get the chance
to teach again soon. There are a ton of young men who could
use my makeup tutorial.

Chapter Thirteen
ABC INTERVIEW

Allow me to give you all a litmus test for your health and well-being. If at any point you find a way to avoid pants with buttons and zippers for more than an entire season, that should be the call for help your body is giving you to say something is drastically wrong.

The winter of 2010–2011 was that time for me. My fill-in opportunities at MSNBC in New York City were not as consistent (I had been filling in pretty regularly for a while; I even did the *Weekend Today* show once). Chicago was starting to feel stale. I had gone through a slew of unsuccessful relationships, moved to what felt like every neighborhood there was in Chicago, yet I couldn't find peace. I now lived in a cave-like studio apartment with just Otis about a block and a half from the NBC tower.

Every day I would work my morning shift, and then go home and drink half a bottle of wine between ten and eleven

A.M. to "help me nap." Then I would wake up, groggy and irritable and leave for work again. During my dinner break, after the 5 P.M. show and before the ten P.M. show, I would polish off that bottle of wine and sometimes open another. This became not only normal but necessary for me to get through my day. I was seeing a therapist, but she was what I call a "yes-therapist"—someone who may as well be your mom, who agrees to everything you say without really challenging you. And that was the season that I stopped wearing real pants entirely. I would go from sweatpants at home, to tights and a skirt at work, back to sweatpants, and once in a blue moon, workout pants (not that I worked out; they just sort of got me thinking about working out). I don't know what it is about pants that keeps me in line, but without them, I live in a fantasy land where wine is calorie-free and a valid member of the fruit food group. I don't eat much when I'm depressed, so the weight gain wasn't terrible, but I was bloated and my eyes were bloodshot all the time. Not a great camera-ready look. Our weekend makeup artist, Diane, knew me well enough that she finally asked if I had been drinking when I got to work. At two P.M. I am sure I reeked of wine. I would make up an excuse, like *Oh yeah, I was just catching brunch with friends.* Brunch with friends was, of course, my solo bottle of wine in my dark studio apartment watching reruns of *Beverly Hills 90210* on SOAPnet and staring at Otis. But sure, brunch.

This went on until around March, when I realized my

contract was up in less than six months. I began to feel the pressure. Where would I go? What would I do next? I had no answers; I only knew that I felt completely worthless on both a professional and personal level. Even if WMAQ wanted to keep me, I couldn't keep doing the job I was hired to do. The five years of double shifts on the weekend and long stretches of being the low man on the totem pole needed to end. The fear of being out of work, or stuck there, forced me to face the fact that I needed to find a way out of this Cabernet-soaked depression. I had now been in television for almost a decade, and I felt like the next move had to be a big one. So I made a call to Rick Ramage.

Rick was a guy who found me way back when I was in my first job in Flint. At that time, he had told me, "You don't need me at this level, but stay in touch and let me know when you could use my help." Every time I got a new job he would call to congratulate me. When I negotiated my own contract in Chicago without hiring him I think he probably thought that was a mistake. This time I knew I needed his help. I told him I wanted to make a big move, and he agreed to represent me. I updated my highlight reel, cut back on my red wine, and put on a pair of pants. None of it was comfortable (especially the pants), but it felt like a win.

I was pushing the hardest to get a meeting at NBC for the *Weekend Today* show, because I had just always had that *Today* show goal. Remember my password *todayshow10*? Well, it was

2010. I worked solely at NBC affiliates throughout my ten-year career and grown up watching Matt Lauer and Katie Couric on *Today* on WOOD TV, and I just felt like NBC was home.

Unfortunately, NBC didn't feel the same. Rick said that while they did love me, they already had Janice Huff (whom I adore, the smoothest delivery out there) and Bill Karins filling in on the weekends, and they weren't looking to make it an official job (Janice and Bill do weekdays at WNBC in New York City and MSNBC, respectively). But Rick also had good news. ABC News and CBS News were both impressed with my résumé reel and wanted to meet me.

As with that "meeting" at NBC in Chicago five years prior, television stations, even big networks, will often say they are interested in meeting you "if you are in the area." Translation: *We are not going to pay for you to come for a formal interview. If you are that into it, you make it happen on your dime, and we will be happy to open an hour in our schedules.*

So I made it happen. I made plans to stay with my high school friend Kelley, who had moved to Hopewell Junction, New York, after college, and I booked my flight to NYC.

I took a taxi from LaGuardia and remember coming over the bridge and seeing the skyline of Manhattan. Not since seeing Chicago for the first time had I felt more ready for a new city to be mine. Career-wise I was more than ready, and personally, I needed a shake-up. I went straight to Grand Central and took the train north about an hour to visit Kelley for the weekend. I remember being nervous waiting for Monday to come. These interviews were going to change my life; I could feel it.

On Monday morning, I took the train back to the city super early and went to a salon to get my hair blown out. Hair has always been the thing that I don't feel confident doing myself. I can do my makeup any day of the week, but hair, ugh! I can never make it look right. I stepped out of that salon ready to take on the world and perhaps a Pantene Pro-V commercial.

My first stop was CBS News. Like every other television station I have ever been to, the building itself was dumpy. CBS News sits on Fifty-Seventh Street near Eleventh Avenue. I was disappointed in how drab and unassuming this building was. I think when you are in television, you always expect that next step to be so glamorous. But this looked almost identical to the sad lobby we had at my first job in Clio, Michigan. In fact, I wondered, did they get their carpet from the same factory? Oh, well; the job itself would be exciting, and I decided to focus on how much I wanted my new life to start in Manhattan.

I met with Mary Noonan, who was the director of talent development at CBS. We had a very nice talk, but I was in and out of there in twenty minutes. CBS didn't really do weather, and they definitely were not in the market for a meteorologist. So why had I met them? I can tell you that no meeting is ever a waste, because the same people in this small group move around, and your paths will eventually cross at another station. Mary now works at ABC in our talent department.

Slightly disappointed at the time, I walked slowly (since I was way too early) a few avenues east to ABC's headquarters at Central Park West and Sixty-Sixth Street. At this time, I had no idea how to navigate NYC and was so pleased with myself that I hadn't gotten lost. I checked in early and thought this lobby was at least a little glossier than the one at CBS.

I have come to learn that the look of the office doesn't matter. It is the people inside who really count. And the people inside ABC were about to change my life for the best. I didn't know it yet, but ABC was going to be the place where I was finally able and encouraged to fly.

I took the escalator up to the elevator and things started to look a lot more like CBS. Either way, this was a new space and I was bringing my A game. I can't remember exactly what I wore, although I do remember I had gone against the classic black/navy motif that is always suggested. It wasn't the Pepto-Bismol pink I had worn to my first interview in Flint, but it was still something that said, *Hey, I'm Ginger Zee, and I am different from anyone you've ever met.*

Barbara Fedida is the talent coordinator at ABC and the person I was supposed to meet, but she had an emergency and had to delay our conversation. So her assistant introduced me to Sheila Sitomer from the News Practices department. Poor Sheila. She had no idea who I was or why I was there. But we had a nice conversation. Sheila then passed me off to Tom Cibrowski, who was the executive producer of *Good Morning America* at the time. We also had a nice talk. He dropped me off back upstairs at the waiting area for Barbara. Barbara's assistant apologized, saying Barbara needed more time, and said they had other people they wanted me to meet. I must have met three other executives, and each time I was corralled back into that waiting area. And each time I sat there, the same tall man in a suit walked past.

By the third time—the time I was really waiting for Barbara—he passed by again and asked, "Are you being helped?" I told him yes, and that I was waiting for Barbara Fedida. With the friendliest smile he said, "Oh, you must be Ginger!" I nodded, and he invited me to walk with him. As we twisted and turned through the confusing hallways, we started going in a direction I had not passed through yet in this building. The windows started getting bigger; the offices were no longer cubicles. My mind was racing. Who was this guy? We reached his corner office and I was surrounded by glass and a fantastic view of the city. I turned around and looked at the name on the door. Holy cow, this was Ben Sherwood's office. This was Ben Sherwood standing next to me, inviting me to take a seat. Ben Sherwood, by the way, was *the president* of ABC News.

Ben and I had a brilliant conversation. He was so inquisitive, normal, and inviting. I never would have thought that the president of a gigantic news organization could be so accessible. We talked about what I wanted to do, and he asked what my end goal was. I told him I wanted to make great TV and get a tornado on live television someday. I told him about my storm chasing, and my ability to anchor and report. I told him I thought networks did too much damage chasing and needed to do more forecasting. He made me feel so comfortable, and I was just letting go. After what must have been twenty minutes, Barbara walked in. She looked frazzled; she had obviously had a long day of something else going very wrong. Her perfect complexion and incredibly thick and beautiful dark hair looked very New York City to me. She had gorgeous shoes and was dressed like she had her own personal stylist who only shopped from the front seat of runway shows. This woman was taking that Tamron Hall glamour back in Chicago to a whole other level. I looked at her shoes and did not recognize them as Jessica Simpson (which was totally top of the line for me, and I still totally love, by the way). They were something else. This *place* was something else.

I was jolted from my shoe envy, when Ben's voice boomed, "Hire Ginger. We need her on our team."

Barbara looked a bit taken aback and smiled. I had a feeling at that moment that Ben's impulsiveness was something Barbara had dealt with before. So as excited as I was to be essentially hired on the spot, Barbara's face told me it wasn't going to be that easy.

I didn't care. I walked out with Barbara and remember turning over my shoulder to thank Ben. He was already deep into his computer, probably making someone else's dreams come true. He looked up and I said, "Thank you." I meant it. This man believed in me. He was going to allow me to do what I do best and give me the opportunities I could never have dreamed of.

I walked out of ABC and into Central Park doing cartwheels in my head. I believe I did a literal Toyota jump. (Remember that jump from the commercial where they freeze frame the guy who got a great deal on a Toyota?) *I had gotten the job. On the spot. I was going to be a meteorologist at ABC news.* I was eventually going to be the first woman and one of the only degreed meteorologists to ever make it as a chief meteorologist on a network. I was breaking glass and taking names.

I was going to The Network, people. In TV news, a network is the golden ticket, the top of the mountain, the endgame. I was having an out-of-body experience it was so exciting. Even though Barbara had warned me that there was much more to figure out and I had to stay in Chicago until the fall, I knew this was home and that it would all work out. After I called my mom, I called Rick and we celebrated over the phone as I rushed to the airport.

When NBC heard about the offer from ABC, they offered

some fill-in work on *Weekend Today*. I politely declined. About nine months later, our team at *GMA Weekend* was beating NBC in the ratings; NBC hired a weekend meteorologist.

When I returned to Chicago I wrote Ben Sherwood a thank-you note. But not just any note. First of all, this was a guy that I knew would appreciate more than an "It was such a pleasure meeting you; I really hope I get to be a part of the team," kind of boilerplate thank-you note. No, Ben needed a special note. The kind of note only a real natural disaster could write.

Ben,

Thank you so much for taking the time to meet with me last week. I haven't had such a fated feeling since my epic adventure to The Price Is Right *in 2003. I know this may seem odd, but stick with me.*

I had been working almost non-stop at my first full-time job as a meteorologist in Flint, Michigan[,] when one morning I woke up and had an overwhelming feeling that Bob Barker was going to die (I am obviously much better at forecasting weather than life expectancy as Mr. Barker is still alive and well). So, I took my first real vacation time and immediately booked a trip to Los Angeles so I could attend The Price Is Right. *I was an independent woman, like Destiny's Child had inspired, so this was a solo, girl-power trip. I was told by a friend that you can*

just show up and people will have tickets to The Price
Is Right. He was right. I showed up early and saw a
handsome man across the street. He was waving me over. I
crossed and he said, "Hey gorgeous, do you need a ticket?"
Then, his friends came up on the sides of him wearing
Central Michigan University shirts and asked, "Hey, aren't
you that weather chick from NBC?" I corrected them,
"meteorologist," and introduced myself to the beautiful man
who had initially called me over. He introduced himself.
Joe Frost. Joe Frost?

In true Ginger fashion, my married name flashed
before my eyes seconds after we met, and it was the best
yet. Ginger Frost. A meteorologist named Ginger Frost.
Can you imagine?

Joe ended up getting on The Price Is Right, and
I ended up spending the remainder of my solo trip with
Joe and his friends, camping on the beach, swimming,
partying . . . it was such a special, free time. Joe and I
dated for about six months after that until an ill-fated end
to our storybook romance in Ann Arbor at a U of M game.

That aside, I learned to trust my gut. Something
brought me to LA. Now I have that same pull to ABC
News. Please help me make this happen. I assure you, and
Bob Barker, that it will be a great move.

My very best,
Ginger

A bold thank-you note, I realize, but I think Ben did appreciate it. And I would come to appreciate all his correspondence to this day. He has moved up and out, as we knew he would. His talents have no bound. And I will forever be grateful to him for taking this chance. He promised I would never work harder than I would for ABC News. He was right.

Chapter Fourteen
FIXING MYSELF

I picked up *The New Yorker* magazine, then stared at the cartoonish cover, partly in disbelief that I was actually living *in* New York and partly in disbelief that one of the only skyscrapers I had walked into was that of a building that held my new therapist. This wasn't exactly the *Sex and the City* version of New York City I had always envisioned. I hadn't even had a chance to check out some hot new restaurant or store my heels in my oven like Carrie Bradshaw; nope, for me the first place I ventured into was the office of Dr. Scott Wilson.

Dr. Wilson quietly ushered his last patient out and invited me in.

We sat in awkward silence for what felt like a full minute.

I don't know if that was a tactic or if it was really just five seconds, but I ungracefully started the session with my current greatest fear:

"If my new coworkers had any idea . . ." I anxiously kicked my feet.

"They don't and they don't have to," comforted Dr. Wilson. He was assuring me that I could get through my first day at ABC News despite the hell I had been through. The hell I had put myself through for much of my life.

That hell that had culminated in me checking myself into a mental inpatient facility in Manhattan about ten days before I started to work at ABC. That's where I had met Dr. Wilson.

This part of the book may seem to be coming out of nowhere. So, let me explain a bit.

Up to this point you have gathered that I can be hard on myself, that I love self-deprecating humor, and that I have never been perfect. A few people close to me read the first version of this story and their biggest note was that I was "too hard on myself." But that's just it. I am. Always have been, always will be. But recently I've found a way to live that is far less self-destructive. I want to share the journey that got me here, because I feel like there are a lot people who could benefit from hearing that they aren't alone.

Those who read the first draft also said that people have an image of me as "strong, fearless, and brilliant." I was so flattered to hear that and I don't think that image needs to change. I do hope it makes you think that even more.

Despite what you are about to read, please know there are years of stories and experiences I don't have time or don't think are appropriate to share. The purpose of sharing the next few chapters is to let you in to a mental health challenge I had, and to give hope to those who feel like there is none. No matter how bright and shiny, perfect and put together someone may seem on TV, she is a person just like you and me. For anyone who watched me that first decade on television, you were seeing the best representative of me. For three minutes at a clip. Here's a glimpse into what was really happening and how I got through it.

Depression has always been my general diagnosis. And by always, I mean since my suicide attempt.

So the term *depression* was not new to me. What I did know is that in the months leading up to my move to New York my depression was rivaling that postcollege low. The girl in the dark room was back, calling me to crawl in. As low as I was, I had this huge job ahead of me at ABC, and even though that dream job wasn't enough to pull me out of the depression, it was enough to give me motivation to get better. I knew there was a strong, smart woman buried inside somewhere, and that's who I wanted to show up at ABC. That alone was saying something. I wanted to show up. That meant I wanted to live. So I was already one step ahead of the other dark places I had ended up in too many times before.

These were skills I had learned from multiple depression guides to help me better cope with my predicament. I did what I was supposed to do when I started feeling really low. I called some of the most supportive and understanding people in my life, such as my cousin Tammy and my mom, told them what was going on—the real ugly truth—and asked for their help.

Collectively, we made the decision that instead of moving into my new apartment on the Upper West Side of Manhattan, I would get to New York and check myself into an inpatient therapy hospital for a week. The place we chose had a concentrated intense program that would separate me from the outside world for a little while and remove any possibilities of self-destruction. It felt like the right decision. But when I got there, I was struck by how cold, scary, and eerily reminiscent of *One Flew Over the Cuckoo's Nest* it felt. The moment I turned over all my personal belongings, including my phone, and said goodbye to Tammy, who'd flown in just to drive me there, I was overcome with a huge wave of panic. I wanted out. Now. This was a mistake. This was not for me. These people were so much worse off than I was. They looked like real zombies. These were people with *real* mental illness, not just my garden-variety depression. The voice in my head screamed, *I don't belong here! What in the world did I just do?*

But then haunting images from my past flashed before me like a movie—shivering, trying to muffle my cries under a table as I hid from my emotionally abusive ex-boyfriend John; my makeup artist Diane asking me, "Have you been drinking?" at two in the afternoon; and all those long, dark winter days I'd spent holed up in my apartment liquored up and depressed and considering suicide. And then I remembered what my cousin Tammy had said when she dropped me off: "You need to do this. Because we don't want to lose you forever."

I will never forget how serious she was, how scared she was of losing me, and I decided then and there that I could do this. If not for me, for Tammy and my mom. They had both seen me go too far down this road and not get the help I needed. This time we weren't taking any chances. They deserved it even if I didn't fully believe I did. So I went into that cell—I mean, room—resolved and determined to face my demons head-on and find a new way to live as a natural disaster that wasn't so exhausting and self-destructive. I know I can't change who I am. I will always have elements of my personality that are impulsive, chaotic, and dramatic, but I had to find some peace in my life as well. I had to figure out a way to exist in which I could let go of the bedlam I'd been addicted to, and learn to accept peace as a possibility—and maybe even accept love. But first I had to learn that I deserved peace and love. Because as strange as this may sound, I don't think I came into this world ever feeling that I deserved either.

As I got older, obviously I took all of those same issues into my relationships with men. Their feelings became my feelings. Joe's disappointment and sadness over the breakup of our engagement became my guilt and shame that I punished myself with for almost a decade. And that's what I did. I punished myself with men.

And in turn I punished them. I was never fair, rarely honest with myself or them, and I was always looking for a void to be filled. Pretty common, but it doesn't always happen in such an unfortunate manner. I was ready to figure this all out once and for all and heal wounds that had been with me for years or even a lifetime.

The first morning of rehab, the hard work began. I got up and easily beat my roommate to the shower. She hadn't said a word to me since I arrived. She was obviously in some serious distress. But even knowing that, I took it personally. I wanted this stranger, this mentally ill stranger, to like me. That's what I sat down and told the doctor right away in the hospital.

That doctor was Dr. Wilson, and he was unlike any therapist I had ever had before. He was distant, clinical, and matter-of-fact. He seemed like he was in his late thirties or early forties and was handsome in a Jason Bateman kind of way—same dark hair and stature. There was no parental warmth; he wanted to get to the bottom of what my issue was and get me out of here. That's what I wanted, too, so I felt good with him right away.

On day one, this guy figured out what at least six therapists in my past had never been able to capture. Or maybe I had just never been able to hear it.

After expressing my distaste for my roommate ignoring me, I dove right in. I started with the event that I had been playing over and over in my head since it happened.

It was three A.M. when I dialed 911.

"What's your emergency?"

"I need help. I need to get away from him."

"Ma'am what's your address?"

I answered. She hung up. And I prayed.

My arms curled around my knees, my body tucked in a ball, I sat, silently sobbing, rocking, just hoping the police would arrive soon to rescue me. I was hiding under a table at a remote resort—hiding from my boyfriend, John.

Until this point, I'd been quietly suffering under the weight of his emotional and verbal abuse, hiding my head in the sand as if it didn't matter because I loved him. But this was different. I was afraid he was going to physically hurt me. Check that. I wasn't afraid he would hit me, I actually used to ask him to hit me because I felt like it would be easier than what he usually put me through. What made this night different is that I truly couldn't take it any longer and needed help to get away. While I waited for the police to arrive, one thought grew as clear as my name: *You have to get out. Everything is at stake. Your career, your emotional well-being, your sanity.*

So, how the hell did I get under this table?

How did a woman who worked so hard to grow up, to be

successful, to learn to make better choices, to end an engagement, to give up a fancy apartment, to mentor young meteorologists get here? How did a woman who always tried to find the self-deprecating humor in being a natural disaster find herself fearing for her life?

I always felt like my life was split in two. There was the successful, happy-go-lucky "wake up and watch Ginger Zee" side. And then the dark, despondent, tortured, and introverted Ginger Zuidgeest. There are times I would look in the mirror and have no idea who the person was staring back at me. I had gotten into these appalling excuses for romantic relationships over and over and I couldn't get out.

"I wish I could erase everything and start over," I told Dr. Wilson.

"With the situation at the resort or in general?"

"Everything. I want a fresh slate."

I knew what came next, and I hated this part—the part where you have to go back and look at your childhood. It was like having your first date over and over, but this was with a therapist. And I had done this so many times. What a waste I kept thinking. But this time I had that glimmer of hope and wanted to get better.

I looked up at him in my hospital-issued giant gown, slouchy socks, and disheveled hair ready to unleash the fury that I believed to be my life and he cut me off.

"Ginger, have you ever seen a therapist before?"

I've probably seen a half dozen therapists over the course of my life. I'm like a therapy connoisseur for goodness' sake. I saw a therapist after my parents divorced, then again after my mom figured out I was suffering from anorexia nervosa—a disease that is rooted in a need for control in the face of chaos, which was a big part of my childhood. For me, the inability to control it all, which of course nobody can, gets turned into self-hate. My whole life, whenever something went wrong, I thought it was my fault. As I grew up, when other people (like boyfriends or bosses, or friends—or even the guy bagging my groceries) got angry, frustrated, or sad, I somehow figured it was my fault and I had to make it better. Or I gave up. And I let the girl in the dark room consume me like I had right after college.

I hadn't felt like that again until that moment hiding under that table and waiting for the police. I felt weak, worthless, and finally aware that drinking more than a bottle of wine a day wasn't going to be enough to erase the memories of John nor the feelings of failure I had now.

I was almost worse than suicidal; I was numb and had been for quite some time. And this zombie was not going to keep this unhealthy lifestyle up at her dream job in New York City. The current issue was the issue I live with always: doing anything and everything to maintain perfection. And above all else, DO NOT FAIL.

In my mind, even with this great job on the horizon, I had

failed miserably in my personal life. Almost everyone I dated after Joe I turned into a form of emotional arsenic. I couldn't turn any of them into the man I wanted them to be and I couldn't get any of them to love me the way I had fantasized about. I had failed. And when I fail I am worthless. Feeling worthless is like feeling you don't deserve to be alive, to occupy space or even breathe. . . .

By the way, that was just day one with Dr. Wilson.

He encouraged me to slow down. I tend to get animated when describing my near suicidal thoughts. But I swear I didn't get here on imbalanced chemicals and bad relationship choices alone. It took me nearly thirty years of chaos to get to this place. That was a big part of days two and three for Dr. Wilson and me to figure out. Here's what we came up with: To be a natural disaster, you need to grow up in some turmoil. You need to find comfort in all things uncomfortable. Which I do. But in my particular case, I am a natural disaster who wants . . . no, needs . . . everyone around me to be happy. I had never attempted to make peace with peace because the need for chaos is buried so deep inside of me, it feels like it's in my DNA.

He then asked me about childhood.

Now, before we begin, I need to make the point that my parents are both loving, supportive, and awesome. As the cliché goes, they both did the best they could. But whatever judge came up with our custody agreement after the divorce should probably be asked to hang up his robe. We spent one day a week and every other weekend with my dad from Easter to Christmas. Then on Christmas Day, we would move all our things to my dad's house and spend all of our time there, except one day a week and every other weekend, for which we flipped to my mom's house. On Easter, we would move again. This was my childhood. While I was grateful for my ability to adapt when I went to college, it definitely set me up for a life where I am only comfortable when I am on the move. For a long time, I really didn't like being in one place. Dr. Wilson and I discussed my need to travel, and to move apartments (I moved four times in five years in Chicago). I was comfortable being constantly displaced. That was home.

Then, there was the chaos.

My mom is a fiery, mostly Italian gal with what we might call a bit of a temper. We all have stories about our parents, and by no means am I blaming everything on the actions of my mom and dad, but there's no way to deny that they had an influence in shaping my personality.

After the divorce, my mom had many single-parent moments that set her off. Whether it was forgetting her purse at the bank or finding our dirty dishes in the sink instead of the dishwasher, it didn't have to be much; but when you are under a lot of pressure as a single mom working twenty-four- and thirty-six-hour shifts and coming home exhausted, your reactions may be slightly heightened. Especially if you are

Dawn. My mom is a pathological perfectionist. She is the ultimate protector and caregiver. She's the woman you want with you when you are in a hospital needing the best care. She's also the woman who spreads herself so thin that when she isn't perfect, or something isn't perfect around her, look out.

Behold, the story my brother and I like to call "the Fruit Ninja."

It was late on a weekday night. My brother Sean and I had school the next morning, but we had nothing to eat in the house. My mom said, "Get in the car. Let's go get groceries." Awesome. We had a blast running the aisles and picking out what we wanted for lunches and breakfasts. I could tell my mom was in a good mood, because we were all singing and Sean and I were even allowed to pick out a few treats. The joy and celebration of full paper bags packed in our station wagon was almost too good to be true. As I shut the car door behind me and let out one more yelp finishing the song we had been singing, I felt the shift before I even turned to see what was happening. The mood had changed. The energy had gone sour. Sean was silent, too. We both watched as my mom fumbled around in her purse. When we arrived home, she started small, murmuring:

"I must have put them in here somewhere. No way do I not have the house keys. No fucking way."

The volume started increasing, and her voice grew shriller. She slammed her purse to the ground.

"No fucking way. Fuck you."

She checked the console in the car, the glove compartment. *Slam. Slam.*

"Fuck everything. Fuck fuck fuck!!!"

"Argh slkdfhaipfhnal,kajlelkfjdkvjsldkj."

I actually don't even know how to type some of the sounds that come out of my mother. But they weren't pretty. This was not the first time we had seen her melt down. It had been commonplace since she and my dad divorced. And my mom had been like this since she was a child. They used to call her Betty Boop when she was a kid, because like the cartoon character, she would count while the anger escalated. As a child, I don't remember many blowups before the divorce, although I am sure she had some. It's just that they became much more common under the extreme stress and pressure of the divorce.

By this point, Sean and I had been through these episodes enough times to know what was coming next. My mom has always had a thing about taking out her frustration and anger on inanimate objects like drawers, mirrors, and doors. It wasn't us, ever. But nothing around us was safe. What she did next went down in the family history books as my favorite blowups.

First she kicked the back of the car and ripped open the trunk of the station wagon.

I couldn't see exactly what she was doing because I was trying to distance myself. (She never hit us, I will emphasize again, but it was always good to give her some physical distance in these circumstances.)

And then I saw it. A plum from our spree at the grocery store, splattered against the wall of the garage.

Bam!

Another plum bit the dust. The soft, ripe fruit I was hoping to have for lunch was smashed against the wall.

Then came the peaches, the apples, and the corn. In fact, almost all the groceries were in her sights. With every launch by this five-foot two-inch dynamo, there were words accompanying the action that you couldn't use on a cable TV show. This was an adult woman who couldn't find her keys and thought it best to hurl produce at a locked door, as if that would help.

Sean and I knew better than to laugh, even though sometimes these fits of rage got to be just downright hilarious. Years later we did laugh, and I still bring this one and a few other classics up with my mom. I know she isn't proud of how far off the handle she would fly, but that's who she was. And now she's much mellower. And medicated. Thank goodness.

I don't even remember how we ended up getting in the house, but I do remember her apologizing for her outburst, as she always did, and then explaining why she was so frustrated and that it had nothing to do with us. And although I knew that, there was nothing in my soul I wanted more than to go back in time a half hour and be able to have those keys so my mom never had to get that mad.

I've had a lot of time to think about and talk about my childhood and my parents, and the best that I can come up with is that they were an unusual pair. As demonstrative and emotional as my mom is, my dad holds back on his feelings and can be critical. Being their kid was confusing. On the one hand, I never wanted to yell like my mom, but on the other, I was also desperate to get some kind of reaction from my dad. In the end, it was a perfect recipe for a people pleaser who thrives on chaos.

As the older sibling, I wanted Sean and me to be safe and happy. I came to a decision very early on in life, the totally subconscious kind we all do as kids, that would affect me until my early thirties. I thought that the only way to be safe was to be perfect. I couldn't control my mom's behavior or get what I wanted from my dad, but I could be the perfect child—perfect, perfect, perfect so that no one could get angry and everybody would love me. Great plan, right? No wonder I was so ambitious in my career; no wonder somebody like John came along who could always keep me guessing about how he felt, and who knew just how to push my buttons to keep me under his control.

My parents both remarried, and starting when I was fourteen, there was a new baby for each of the next three years.

First my dad had my half-sister Bridget.

Then my mom had my half-sister Adrianna.

And my dad had my half-brother Walter.

My parents had both started new families. And yes, we were always part of the family. But now that I have my own child and know how much effort and love a baby takes, I can't imagine having a fifteen-year-old to deal with at the same time.

We were most definitely still loved to the moon and back, but life changed when those babies came into the picture. My parents were late to pick me up from sports practice; neither of them made it to all the school events, field trips, or cheer competitions. I think I felt like we had taken a bit of a backseat to their new families.

So we did what any child would do. We both reached for a microphone. My brother is the lead singer of the band, The Outer Vibe. I went into television. We both took on professions that demand attention.

It's strange because now that I can step back and have the wisdom to see it differently, I was almost trapped in that sensitive teenage mind-set where everything revolves around me. That was not always in a bad way, but in the worst way for me, in that I would absorb other people's feelings. I didn't emotionally mature.

That's what got me into so much trouble, and it was entirely my fault for not asking more questions and communicating more. I couldn't separate my feelings from anyone else's.

And that's exactly what Dr. Wilson helped me untangle during our therapy sessions. I can't believe it took me that long in life to figure it out, but I'm grateful I got to him when I did. The first thing Dr. Wilson explained to me—and it was a life-altering revelation to even hear this—was that everyone is entitled to their feelings, but it is unfair to absorb other people's feelings. Wow. So somebody is mad and it's not my fault? Conversely, somebody is pleased and it's still not because of me. Life changing. But it's not enough to just know this. In order to practice it, he taught me to put up an invisible fence. He taught me to recognize and acknowledge how the other person is feeling, but then block it from jumping into my space. He told me, "When confronted with another person's actions in response to their feelings, say, 'I am sorry you feel that way.' You can genuinely feel empathy or compassion without absorbing all their negativity. Especially when it has nothing to do with you." This idea has revolutionized the way I look at the world, and I don't think it's an exaggeration to say that working this tool is the foundation for everything I am so grateful for in my life today, including my job, my husband, my son, and my happiness.

It's hard not to look back at my younger self and wish I had known about this tool as a kid. If only I could have put up a fence when my mom was having an eruption, I might have had some peace, I might not have gotten addicted to the drama and the way it made me feel. When I wasn't sure what my dad was thinking and I made up the most negative emotion I could imagine him having about me, I definitely could have used that fence. But it's wishful thinking. Sometimes I think it's a miracle we all survive our childhoods.

Chapter Fifteen
CRACKING MY CODE

"**H**e would have been abusive even if you didn't cheat."

Dr. Wilson and I were on day four, going back through my failed relationships and the one that I couldn't shake. The one where I had literally failed. Miserably. Or had I?

On a cross-country trip to Los Angeles recently, I watched *The Girl on the Train*. I'm not sure if you do this, but I had one of those horribly embarrassing moments where the movie touched me so deeply I started bawling with no concern that I was in a capsule filled with strangers. I had read this book but did not remember the end. Spoiler alert: in this book and movie the main female character is painted as an alcoholic disaster who ruined her marriage and has an unhealthy obsession with her ex-husband and his current wife. In the end, you learn that *he* was really the monster. She was fallible, yes. She was drinking, yes. But they went through and told the story quickly again of their marriage failing, of her obsession with the ex-wife . . . from the perspective of him controlling and manipulating her.

I had been the girl on the train.

I had been controlled and manipulated.

And while I did make several mistakes, he, John, was really the monster.

Writing this book has been so eye-opening for me. I beat myself up for years because of my mistakes. But every story has two sides and the truth. I know my side and I have worked hard to find the truth.

Anyone who has been in a manipulative and controlling, emotionally abusive relationship will find this description eerily reminiscent. Because they are all the same. Let me just say this right now—just in case anyone reading this is in an emotionally abusive relationship but think their love is so strong that it is going to get better and it will be worth it. It won't. It can't. I wish I had something different to say. I wish I would have understood sooner. It is not healthy nor normal for your partner to isolate you from family and friends, to constantly have all your passwords and read all your texts and e-mails. No matter what your situation, you should be loved and your partnership should be supportive without drama.

When John and I met, I was actually living with my

boyfriend Angelo. Angelo was tall, dark, and very funny. He also had a seasonal job. I met him in his season of work; by the time fall came around and the weather turned, Angelo moved into my apartment with me, essentially out of work until the next spring. Suddenly, I had a barely employed boyfriend, which if I'm being honest, I hadn't given much thought to.

Angelo spent his days playing video games and taking long walks in the city. He was helpful with my dog, Otis, and I appreciated that but as we headed into winter, it felt as though he were hibernating. I knew this was just part of Angelo's schedule, year in and year out, but it was new to me having a boyfriend who didn't change out of his sweatpants all day, and as much as I didn't want it to bother me, it did.

Making matters worse, I was growing restless at work. It was beginning to feel stale, doing the same exact forecast from the same exact studio every weekend. I was starting to resent the "normal" people who were out in the world, eating and laughing and having a life while I was stuck inside a dark studio for sixteen-hour days every weekend. That entire winter, I focused on finding something to make my career advance and let my relationship with Angelo fall to the background. I was far from happy.

And then spring arrived. The flowers bloomed, the sun was shining, and a Facebook message appeared from a guy named John. He was interested in offering me an opportunity that would help my career. We scheduled a meeting and I was immediately smitten. John and I spent four days working on a project together. The last night neither of us could deny the chemistry any longer. We kissed and I heard angels sing.

This was my future. Everything felt right. He smelled right. Do you know what I am saying? His skin felt and looked like the skin I wanted to be next to forever. There was a magnetism between us that I now believe was there to teach me a lesson in how to find value in myself and not allow someone else to determine my worth.

I went home and told Angelo we had to break up and he had to move out. He was not pleased and it wasn't easy, but I knew I had to do it. I had learned this lesson before. When something doesn't feel right, get out. Or you'll end up almost getting married. And I was not going to catch up with Julia Roberts's numbers. This time as I ended it with Angelo I felt this sense of freedom I hadn't ever felt in my life, even when I'd ended my engagement, which was so loaded with guilt about the pain I'd caused my fiancé. This time, I was ready to move on and get out of the quicksand of stagnation.

John was the polar opposite of Angelo. Like me, he hated sitting still. We made plans for him to come visit me. I stocked my refrigerator, I cleaned my apartment, and I couldn't wait for him to meet my dog and my friends. I had never been a long-distance dater, so I was naively optimistic.

Just a few hours before he was supposed to arrive, John called and had bad news. He wasn't going to make it.

I was pissed.

I'd broken up with my boyfriend for this guy. I'd turned my life upside down for this guy. I'd bought barbecue potato chips for this guy!

But I'm forgiving. I was willing to give it another shot. I went once more to see John on his turf, and we had a fantastic time together. I felt myself falling for him more and more as I got to know him better. I was seeing his silly side, his caring side, and I could feel his guard dropping. We made plans for him to come visit me . . . again.

And again, he didn't show up.

I didn't get it. Wasn't I important enough? Wasn't there more to life than work? Didn't he feel my skin was right, too?

If I could broadcast this to all women (or men) thinking about changing anyone while they are dating, please listen: *You must listen. People tell you who they are up front. You just have to be willing to listen and then accept it.* But of course, that's the thing right? As a natural disaster, my fantasies are critical to my survival. They are, in fact, my reality. So why would I pay attention to reality? That's no fun.

When I met John, he let me know in our first conversation together. He said something along the lines of, "I have never been as interested in a relationship than I have been with my work." He said that to me! Clearly. And then I decided to ignore it. Why would I pay attention to such an obvious red flag? Why not choose to process this information as evidence of how much passion he had for his work—which was something I also had, so wow, that just confirmed we were meant to be. Why would I ever choose to read that as a way of a man telling the woman he has just gotten involved with, *Hey, don't ever get too attached to me because I love what I do more than I love people, and that will at some point down the line, include you?* Again, I wasn't interested in reality. Reality is very annoying, inconvenient, and distracting.

So instead I buried all the evidence. The comment, the cancellations of his visits, anything that got in the way of my fantasy. Finally, after the two aborted trips, John came to see me. I once again cleaned my apartment, groomed Otis, and picked John up at the airport. But as soon as he took one step into my apartment, he opened his laptop, and he made it very clear that this relationship was costing him critical time away from his work. Within an hour, he had taken jabs at my friends and family and even started challenging me on some of my religious beliefs and values. I thought that was odd and chalked it up to him being in a bad mood from traveling.

That night, I had made plans for my best friends to meet John. Just as we were leaving, he told me he needed to stay back and work and just didn't feel comfortable because of his social anxiety. He went on to explain that he could never be a "social butterfly" like me. He is one of the only people that has ever made that phrase sound like a bad thing. I didn't know why at the time, but I was already changing, morphing to be what he wanted me to be. His manipulation was heavy; his ability to vilify others around me and isolate me from loved ones was masterful. Despite all that, I kept telling myself I wanted, no, needed, for this to work.

What happened next still haunts me. I can see it now, and in retrospect, with a lot of therapy, I can understand it. Cheating on John was something I did to protect myself. Somewhere deep inside I must have known that John was bad for me, dangerous even, but without the courage to leave, I would have to find a way to make it implode so he would leave first. I was also afraid of being alone, so cheating on John was a way of hedging my bets. Plus, as a natural disaster, I thrive on turmoil.

I got a text from my now ex-boyfriend Angelo and I texted him back. I shouldn't have, but I did. I'd made the break with Angelo; it was over. But John wasn't who I thought he was going to be. So, I returned the text.

Angelo wanted to get together for coffee and a talk. Nothing more, he said. He just wanted to know why, why had I given up on us? I said yes. It seemed harmless enough. I owed him that, right? We met, we talked, and I went home. That should have been the end of it. But I was lonely. That friendly coffee soon led to a friendly lunch later in the week, which led to a friendly dinner a few days later and eventually led to friendly sex. I know, right now you must be judging me hard. But Angelo was warm and available, safe and supportive. Everything that John was not. Angelo was edging back in and filling the void John left open. Angelo truly loved me. He built me up without feeling the need to tear me down. All of this should have made me question why I was involved with somebody who didn't make me feel safe, but it didn't, not yet.

The morning after Angelo and I slept together, I felt terrible. So guilty, so ashamed. It was so stupid, so unnecessary. I tried to justify to myself why I hadn't done anything wrong. John and I had only known each other for six weeks. At one point, I even tried convincing myself that I hadn't had sex with Angelo. But here's the thing about your gut—it doesn't lie, and it will not be lied to. You can ignore it, but you will pay the price. The thought of telling John and coming clean was never an option. The rest of that summer I danced between the two guys and dated other men, acting as if I were not committed to anyone—not once being honest with myself or them. It wasn't always physical, but emotionally I was all over the place. I fantasized about how I could mesh Angelo and John into one to get the perfect man. If you ever find yourself saying this, please exit all current romantic relationships. That means that neither is right.

Late that summer, John and I went away together. To me

it was the trip that would finally determine whether John and I should really be together. It was a fantastic trip. We stayed in a swanky hotel and hiked. I finally believed we were falling in love. As a natural disaster, I felt like if the trip went well, we'd be okay. It would be a sign, and I would just forget I'd cheated on him. I know now that traveling is the *worst* metric for determining the health of a relationship. I know some couples who are terrible when they travel together, and awesome at the mundane details of their regular lives. Then, of course, there are couples like John and me (usually in the early stages of dating) who are great on vacation and a catastrophe back in reality. Four days into our trip, John and I were having such a great time, we decided to extend our trip. After rebooking our return, I jumped in the shower to get ready for a romantic dinner.

What happened next would shatter my life. From that point on I would be a shell of myself attempting to make up for the wrongs I had done. Wrapped in nothing but a towel, I walked back into the bedroom and found John just staring at me, holding my phone, a text messge bright on the screen.

"Who's Angelo?"

I immediately felt vomit rise in my throat. My heart raced and I thought I would pass out. My immediate reaction was culled from every cop show I've ever seen. Deny, deny, deny.

For the next hour, I talked around myself in circles. I hated lying to him, hurting him, but I couldn't stop. When

I broke off my engagement, I had also been *one of those people who hurts people*, but this was worse. Ending my engagement had been the right thing to do for me and my fiancé, but cheating on John was wrong in every way and I knew it. My gut was smirking in the background. *See, I told you so.*

John didn't need me to explain anything. He'd figured it all out himself and decided that I was playing my ex against him. My lying and denying had worked, or so I thought, and we decided to go to dinner.

When we got back to our room after a tense meal, I went to sleep. John said he was going to stay up to work.

I woke up in the middle of the night and found John sitting in the glow of his laptop just staring at me. He demanded more answers.

"Tell me the truth. Have you seen Angelo since you broke up with him?"

"No."

"I don't believe you. I am going to assume you have seen him. Have you kissed him? Slept with him?"

His questions kept jumping to the next level even though I was doing my best to contradict everything he was saying.

I just kept lying. It made no sense. I was still desperately trying to save a relationship with a guy who consistently made me feel small, who clearly told me he loved his career more than people, a guy who didn't really want to become a part of my life but wanted me to be his and only his when it worked for him. This was a guy who had made it painfully obvious that I was an inconvenience.

I had done John a huge disservice by casting him in the role of my knight in shining armor who took me on great adventures and rescued me from the banality of a weekend studio job and a boyfriend who was addicted to video games. That wasn't fair to John, Angelo, or especially me. We need to learn to be our own knights in shining armor, and when we cast others in that role, the universe seems hell-bent on proving us wrong. That's what I was living through at that moment.

I woke up a few hours later and John was still glued to his computer, this time with a new development. John had messaged Angelo on Facebook and they were sending messages back and forth to each other.

"You have one more chance, Ginger. Did you sleep with Angelo?"

I sat there frozen. I was exposed. I could see Angelo's profile picture across the room. John knew everything. That has to be every cheater's nightmare. They were messaging each other with time lines, comparing notes on my stories. The jig was up.

Once he had his confirmation, John started calling me the nastiest things I have ever been labeled. He attacked my family, my friends. He insisted I hand over my phone so he could read every text message not only between Angelo and me but between *everyone* and me. It was a horror show. He then forced me to give him my password to my e-mail. He was hurt and now no one was safe. He didn't let me sleep; he just kept hammering me with questions, making me admit what a horrible human I was. I couldn't stop crying and couldn't stop puking. I was so upset that my body was revolting. He went on for hours, repeating the same points until he got me to answer the way he wanted.

I was now pacing on the balcony of our room; this torture and cross-examination were seemingly endless. Wasn't he satisfied with my admission? Why did he need so many more details?

I had never been in this position before, so it was a very odd place to find myself in, and I couldn't tell him that this was going too far. By eight A.M. I was exhausted. John hadn't slept a wink, and hadn't let me, either. I would rather have gone to jail for a night than have had to endure this.

I felt so relieved when I was "allowed" to leave the hotel room. The whole night I had felt imprisoned, partly by my own guilt, but mostly by him. This was almost the perfect outcome for an abuser. To have his target make such a glaring mistake meant he would never have to keep me down going forward. I'd do that to myself. He could just emotionally "kick me" whenever he wanted. We packed our things, drove to the airport, and flew home.

I was destroyed. I hadn't slept now in almost two days. I

had hardly eaten. My eyes were swollen like giant pillows from crying. The accusations that were thrown at me were more than I could handle. I'm not saying I didn't deserve it, but as a lifelong people pleaser, this was my ultimate nightmare. But I deserved it, right?

My mom picked me up at the airport and I figured John and I were over. It should have been over. Our relationship was obviously shattered. At any other point in my life, with any other person, I would have left. I should have left. But somehow, we kept going. Within hours of getting home, John told me that despite my mistakes, he didn't want to live without me. I was his "everything." What I really became was his emotional punching bag.

From there on out, John pulled all the strings. He needed to know my every move at all times. I validated every story of where I was, what I was doing, and whom I was with, complete with photos and GPS markings. With one phone call, I would abandon everything and everyone in my life to get to John. All I wanted was a chance to prove that I could be the best partner anyone could ask for. I was so busy waving my white flag that I never saw his blatant red flags.

John's mood would shift dramatically from moment to moment. No matter where we were on the globe, I was always traveling with Jekyll, Hyde, and at least three other characters.

One of those characters loved control. He loved to interrogate me, and once he found a nugget, he would demand I hand over my phone, my e-mail passwords, everything. And then he would dig deeper than a detective on *SVU*. This guy would find some archived e-mail from ten years earlier that even I had forgotten about and start accusing me of sleeping with whoever had sent it. Some of the accusations were ridiculous: students I had taught, my best friend's husband, my best friend's husband's neighbor.

Every time John and I were together, things began the same way. We were *great* when we were great. But sometime within the latter half of the first day, John's eyes would glaze over. He would turn from a loving and caring person to a cold and frightening monster. He would start prosecuting me. And I would start crying. Then we would fight. And for three days it would be pure hell. He would drop me off, my eyes always swollen from days of crying and trying to make him love me. I would get to a safe place away from him and like clockwork, he would reverse course. Back to nice John. Back to *I love you*. Back to *We will get through this*. Sometimes I would turn my phone off or be in the air traveling, and when I powered back up, it would always be flooded with panicked, desperate texts from John apologizing for his terrible behavior. He would promise that it would never be like that again. But it always was. And every time, it got worse.

Chapter Sixteen
A NEW, BETTER ME

Even today with all my progress and growth, I have to remind myself to separate my feelings from others and to avoid absorbing other people's energy. I've heard it referred to as "staying in your lane" and that makes sense to me. The bottom line is I can only be responsible for my own feelings and actions; the rest of it is just none of my business. Simple idea, hard to execute, and profoundly life changing when it's worked to the best of my ability on a daily basis.

I'm not saying there aren't relationships that can recover from cheating, because I've seen it happen. But John and I were never committed to each other before I slept with Angelo. We lived in different cities and were both wildly ambitious. And he was an abuser. We were doomed. It's wild that an intelligent woman like myself could get wrapped up in that web of manipulation, but for anyone else who has gone through this, remember: it doesn't matter how smart you are. Abuse begets abuse and does not discriminate.

See, that's all my side. I am certain he would tell the story much differently. But the truth I believe has something to do with both of us being a terrible fit for the other. I have this beautiful dream that he is different to other women. That I brought out the monster and there is this generous, kind-hearted man inside. I would often ask myself, when were we doomed? The moment we met, the moment I cheated, the moment he found out? Well before John ever met me? Maybe that's the wrong way of looking at relationships. It's hard, most of the time impossible, to see the big picture when you're in the middle of a storm. I know I certainly couldn't. But looking back, I'm hesitant to use the word "doomed"; if anything, we were fated. And by fated, I don't mean in that Cinderella romantic-comedy kind of way. I mean we were drawn to each other to learn certain lessons from each other. You don't even have to believe in past lives to take this idea in. It's more like there's something intuitive that we pick up from each other that draws us to certain people. And maybe in romantic relationships, all those hormones that make us sexually attracted to each other—and with John, there was no denying or even fighting that attraction—are nature's way of making sure we don't pass each other by. John had a lot to teach me about myself and I'm grateful for the time we spent together.

That's a strange thing to say when you look at the story I started telling Dr. Wilson. Reminder: 911 call.

This was going to be the trip where John treated me well the whole time. That's what I told myself, and what he always promised before every trip.

My friends were throwing me a party. I called John and invited him, but he made up some excuse. I was supposed to leave to meet John for a trip the day after my party.

About forty-eight hours before the party, John called and gave me an ultimatum: fly to him tomorrow morning or he will know that I don't love him and am cheating again with someone else. This was a ludicrous claim of course, a peak of manipulation and craziness, but what I did was even crazier.

I asked if I could wait one day so I could make it to my party and he said no chance. It was tomorrow or never. So I did it. I skipped my own party. And I did it for John.

I still can't bring myself to look at the photo my large group of friends took for me, standing around the cake with my name on it, while I was flying to "Abuseland."

I joined John and in true manic fashion (on both our parts) we hiked as soon as I arrived. It's almost cheesy how the abuse aligned with the ascent and descent of our hike. As we climbed the mountain, John built me up. He talked about getting engaged, and how much he loved me as we climbed higher and higher, the sun shining. It was nearly perfect. As soon as we started our descent, however, he started preemptively accusing me of the make-believe cheating I would do in the future with people I hadn't even met yet, coming at me harder than he ever had before. I couldn't wait to get to the bottom of the mountain so we could be near other people. As soon as the trail opened up, I started to run. John chased after me, throwing insults at me like it was a game of paintball and he was determined to leave me splattered head to toe with hate, pain, and doubt. Things got even worse that night when I received a text from an old college friend, who was just texting to catch up because we hadn't talked in years. John grabbed the phone and read the text as evidence that I was currently cheating on him.

I told him he was wrong, but it didn't matter. It never did. He told me he needed to step out for some air. He slammed the hotel door and left.

In that moment I briefly thought about packing my things and trying to escape before he got back because I knew this night was about to tailspin into a sleepless wreck. Before I had a chance to even get up from the couch, John slammed the door and came back in demanding my e-mail password and started sleuthing. When he found an ancient e-mail from that same college friend from more than ten years before alluding to a brief college hookup, all bets were off. He'd found his "proof." I was doomed.

"You are worthless and no one will ever love you. . . . What a waste of time you have been. Every and any woman is better than you . . . You are so lucky to have had this time with me because now it's over, you lying sack-of-shitty whore. . . . Your dad is an idiot and your sister is . . ."

You get the point. But then he went a step further.

"Give me your phone. I'm going to text this guy and get the truth. Give me your phone," he ordered.

I handed it over. I had nothing to hide. John pretended he was me and started texting him, fishing for something that did not exist. It must have been so strange for my old friend, going from our friendship where we barely talked to this assault of text messages.

When John was satisfied that there was nothing more to learn and he wasn't getting the payoff he wanted, he handed me back my phone with a unique stipulation: "Tell him you no longer want to be friends with him. Tell him he should never contact you again because you are in love with me."

I paused.

At this point I always did everything John said. I skipped parties with friends; I didn't call family back. I was his puppet, because I thought it would help me someday prove that I wasn't a bad person. I had cheated, but I was and could be a great woman.

But by this point I was so beaten down, and now he was

reaching to the extreme edges of my circle of friends. I held the phone tightly.

"*Do it!*" he screamed.

I started typing and did not want to hit SEND. I didn't mean any of those words on the screen. I loved my friend, and I didn't want to lose him. I knew this was just one more step in losing myself.

Any of you reading this who have been in an abusive and manipulative relationship know that isolating your victim is crucial to abuse working. Without support around me, I couldn't stand up to John or recognize his abuse as easily. I was getting wrapped tighter and tighter in his web of control.

"Send it, or we are done."

I hit SEND.

And it still wasn't enough. Of course it wasn't.

He launched into another attack immediately. He started searching through my e-mails again, pausing only to call me a name or ask me a ridiculous question. The words started flying faster than ever, insults to proclamations of love. Everything was always my fault. He kept searching my e-mail, asking more questions. Now it was two A.M., and I just couldn't take it any longer. I couldn't believe I had just sent that text to my friend. Everything started to become clear. I finally saw myself in the mirror. The good me. The strong me. The me that would not take one more moment of this abuse. I had finally had it. I stood up and said, "I can't do this. I'm done." And that was it. It took me so long to find the courage to say just those few words, but I did it.

John looked at me as though the puppy he'd been kicking around had just bitten him. I packed as fast as I could, with John standing over me vacillating between pleas for me to stay and good riddance that I was leaving. I wheeled my bag out to the hallway to the lobby and realized there was no one there. We were in a small resort, and this was not the Holiday Inn. *No problem,* I thought, *I'll just sleep on the lobby couch.* Then the texts from John started coming. They were threatening, then loving, then desperate, begging me to come back. I refused, and of course that made him even angrier. (I wish I had these text exchanges, because I can't remember them exactly, but I know it was bad enough that I decided to hide.)

I ran up to the third floor and scurried around the pool, checking doors that were all locked to find a place to hide from his crazy. I could hear him calling my name from the staircase. I got in the elevator, and went to the second floor and found a table to hide under. It was the kind of cheap hotel table with plastic clips and a Velcro skirt in a loud floral pattern used for banquets.

I called my mom from under that table.

This was not the first time my mom had answered a three A.M. call and heard me crying with John yelling in the background. I whispered, "Mom, I'm scared. John is at it again, but worse than ever."

"Honey, slow down," she said comfortingly.

"I am in the middle of nowhere, there's no one at the desk. I am so scared. . . ." (My cries were muffled; I was still hiding.)

"Ginger. I want you to breathe. Take an inhale and an exhale. When we get off the phone, I want you to call the police so they can help you out of there. You need to get away from him. He is a monster."

And she was right. I knew it. I'd known it for a while. The problem was I felt like a monster as well. I had become a dreadful person who barely recognized herself.

And that's when I called 911.

A few minutes later, a squad car quietly pulled up to the front door of the hotel. I watched from the second-floor window and made a run for it.

I didn't want to see John. I didn't want to see him ever again. And I didn't know what he would do if he saw me actually escaping.

As I turned to get in the passenger seat I realized this was the first cop car I had ever ridden in. The policeman took me to another hotel and I checked in and got a few hours of sleep. I blocked John from all of my forms of correspondence and told myself that was it. I was finally going to get myself back. I would apologize to my friends and family he had forced me to cut ties with, and rebuild all my relationships. I would rebuild myself.

This was one of many stories I was determined to work through in therapy. A compilation of experiences from random points in my life that brought me to this place. My one to two hours with Dr. Wilson every day were the best part of being in that hospital. After our sessions, I went to lunch. They definitely did not encourage social interaction at the hospital, so I had a lot of time to reflect and plan on how I would change the way I lived. In the afternoon there was a group discussion where I gained major perspective. There were others there who had lost everything and everyone in their lives because of drugs, alcohol, or severe mental illness. So my starting point in recovery was gratitude. I was depressed, but I knew I could recover.

While the experience in the hospital was frightening, and every day I was in there I thought about leaving, I kept telling myself how important it was to get better before I started the job I'd wanted all my life. But it was more than just the job. The job I knew I could handle. I also needed to alter my core. I wanted to learn honesty with others, but mostly honesty with myself. I wanted to learn to communicate, stand up, and not avoid confrontation. I just couldn't screw it up this time. I would not let that happen. I was determined to put in the work to give myself the best possible chance to succeed at ABC and in life.

After my week was up and I was released, Dr. Wilson and I made a plan to see each other twice a week. I decided to stop drinking completely for four months. I also started working out five days a week. The funny thing about making good, grown-up choices is that they snowball into more good choices. I drank water and I ate right. It was another full-time job being committed to my health and well-being, but it was more than worth it. I had some help for sure, but it feels good looking back on that time and knowing that I stepped in and saved my own life.

Let me just say this about the drinking, because it's been mentioned a lot in this book. (And I've certainly done more than my fair share of it.) I understand that alcoholism is a disease, and I know several people whose lives have been saved by rehabilitation programs. I did a lot of hard work getting to the bottom of my truths and my issues, and it turned out that for me, drinking was a coping mechanism, not a disease. My disease was depression and self-hate. My addiction was to consuming the emotions of those around me. Once I tackled those demons, the drinking just kind of fell away and lost its value.

This is the first time I have told this story about my recovery and the depths of depression it took before I began to climb out. It feels good to share it and air it out in the open, and I'm a little amazed I had the strength to keep it a secret for so long. There's a saying, "You're only as sick as your secrets," and I can see the truth in that. It's wild to think back on starting at ABC and the fact that my new coworkers had no idea that the young, excited meteorologist before them had just left a mental hospital the week before she stood at the same desk on *Good Morning America*. When they say "people can't change" I disagree. I did. I think of that time in the hospital and the year after with Dr. Wilson as the time I finally grew up and out of my adolescence. I don't think I am immune to depression in the future, but I do think I'm a lot safer now. My one hope is that sharing my story will have some value to somebody who is still carrying their own secrets, because we all have them, and that whoever they are, they find the courage and the strength and love for themselves to get help. None of us gets through this life alone. Certainly not me.

Chapter Seventeen
TRANSITION TO ABC

My mom picked me up from the hospital, and the fierce hug I got from her was everything. She told me how proud she was that I had made it through the week and generously offered to stay as long as I needed her to.

When we arrived at my new apartment, almost all the boxes had arrived from Chicago, and we started moving them in. In the process, I had my back slip and pull out. Have you ever had that? It is the worst pain I've ever experienced. My mom got me comfortable, and I turned on the TV to watch my future coworkers on GMA *Weekend*. They all seemed so great, and I felt a wave of excitement rush over me. Then, when they flipped to the local ABC, a meteorologist named Amy Freeze (her real maiden name) popped up. I had known her when she worked at Fox in Chicago. I texted her and told her she looked great and hoped she was enjoying ABC. We texted back and forth and realized we were actually living in the same building. She asked if I wanted to go work out together. I told her I had pulled my back; coincidentally, her husband was a chiropractor. He worked on my back and fixed me up before I even got to work that week. It made me feel so at home. Take note; Amy Freeze will soon become another important part of my story . . . when it comes to my husband.

The week before my new job began, I had lunch with my new executive producer, Matt Frucci. We met "on campus," as he called, it at PJ Clarke's in Lincoln Center, and I liked him immediately. He was so intelligent and forward-thinking. He listened intently to my ideas for the show. He was so New York! I felt an instant connection, and he assured me I was going to adore my new coanchors.

It's always so funny to me that in TV you rarely get an orientation when you start a new job. Gerard, my weather producer, met me in the lobby on the first morning I would be on air. Gerard had been a meteorologist at ABC for almost thirty years. He is a handsome, fit older gentleman with the gentlest voice and demeanor. I remember when Matt told me about Gerard's being my weather producer, I was skeptical, because I'd always done it all own my own—the graphics, the forecasting, everything. But obviously I wasn't about to ask them to fire someone, and I wanted to be a team player, so I told myself I would just test it out. About a week later I realized how crucial Gerard would be.

That first morning on the job, Gerard led me into his graphics space and we talked about meteorology and nerded out a bit. He gave me a rundown of what he thought we should cover, and I gave him my input. We planned for two weather hits; both were in the show for about two minutes total. Then he walked me down to hair and makeup. What a luxury! I had not regularly had makeup and never had hair in Chicago. Then I was introduced to Katia.

"Good morning. I am Katia. Sit."

Katia, a petite woman with fiery red hair whose Russian accent was still so thick that she immediately reminded me of my oma, my Dutch grandmother who had recently passed. Katia was direct, cynical, skeptical, and full of pessimistic comedy. She is a true artist; she paints your face like she's preparing you for an exhibit at MOMA. We had some good laughs and I felt at ease.

As I was getting made up, my new coanchors stopped by, one by one, to welcome me.

Dan Harris was first. Slight in stature, huge in dry wit and talent, Dan has a twinkle in his eye that just makes him so special. He dug right in, asking where I was from, where I was living now, how I was liking NYC. He told me NYC was the only city in the world, the best city, and that I was going to love it here. He took a funny jab at Ron Claiborne, the news reader for GMA *Weekend*, as he walked in. Ron has been with ABC for thirty-five years; he is the consummate professional but also has a very dry sense of humor. What a wonderful place this was! Everyone was sarcastic and hilarious, and self-deprecating like me. Ron and I chatted, and I felt great! Two down . . . now just the woman coanchor to go. As I had learned through several catty experiences in television news, women can be weird. I had all my antennae on high alert. And there she was: Bianna Golodryga. Bianna, also Russian, gave a good morning in Russian to Katia, then leaned in and hugged me. Right off the bat! She was so stunning! The most beautiful pout, high cheekbones, flowing gorgeous hair, and gigantic perfect eyeballs. She was playful and warm and everything that I didn't expect.

And finally there was "Fonzie." Alfonso Pena is his real name, and he's the best stage manager in the business. A stage manager is a crucial part of a successful show; he's not only the person who counts you down, but he is the direct connection from the anchor to the viewer. Fonzie isn't just great at his job, but he also has a way of making the folks on TV look their best and feel even better. He's a legitimate comedian and always knows how to lighten the mood. In my opinion, he is one of the best parts of ABC News, bar none. Whenever I would have a low moment, Fonzie was the one that kept me going. There were many days in my first months when I would finish a broadcast and his humor or kind words would deter me from wanting to drink or self-destruct. Thank you, Fonzie.

I went on the air with that team, and I felt like I was with family from the first moment. We all exchanged phone numbers and had lunch that week. It was so inclusive, and they all had such amazing experiences and life lessons to share. I was the newbie, but they already respected me. I will say right here and now that some of my best on-air moments at ABC are still with that team. Dan Harris is the ultimate in generosity. He finds your strength, makes you look good, and makes you look like part of the team. And in turn it makes him look great. It is a skill that few can achieve. He is heady and introspective on television, and it works for him without ever feeling pretentious. His vocabulary is annoyingly . . . huge. Mine is obviously not.

Matt (our executive producer) sat me down a few weeks after I started and said he wanted to get me out in the field for more than just storms. I was definitely down with that idea. He said he had been trying for years for one of his anchors to go paragliding. I had to look it up, too; don't worry. Everyone always thinks it's parasailing, the one where they pull you with a large parachute-looking thing behind a boat by a rope. Paragliding does not equal parasailing. Paragliding is the one where you actually run off the side of a mountain and fly, free of any boats or ropes. Parasail, paraglide, hang glide . . . I didn't care. I loved that I was going to literally and figuratively spread my wings and fly.

Matt told me that he had a producer and shooter, Niels Dachler, who was connected to the Paragliding World Cup. That year it was in Valle de Bravo, Mexico. While I was there, he said, I could shoot a story about monarch butterfly migration and the very serious threat of their extinction. That way he could justify getting two stories out of one budget. This all sounded right up my alley. If I had to jump off a mountain to do it, I would, without question.

I traveled to Mexico with another producer named Rich McHugh. We flew to Mexico City and met the "fixer," who would drive us three hours into the mountains southwest of Mexico City. A fixer is usually a person who can help you translate, navigate, and smooth your entry to areas that are not frequented by tourists. They help coordinate interviews and make sure you get treated well and not taken advantage of. Our fixer turned out to be two people, a couple in their midforties who didn't speak English. I had a minor in Spanish, but it had been years since I had used it, so there wasn't much communication. That was fine; we settled in because we knew it was going to be a long drive. As the trip closed in on four hours and we continued winding through the mountains, the road started to dangerously narrow, and the formidable cliffs around us plunged us into darkness. Rich and I kept staring at each other. We would type on our notepads instead of speaking just in case our drivers, in our minds now captors, could hear us. I typed first.

I'M A LITTLE SCARED.

He responded, ME TOO. DO YOU THINK THEY ARE GOING TO KILL US?

I responded, I DON'T KNOW. LET ME TRY TO ASK HOW MUCH LONGER.

"*Diez minutos,*" the woman responded.

Okay, ten minutes . . . but didn't they say that ten minutes ago? Just as I was about to figure out a way to open the sliding door and duck and roll out of the moving vehicle, one of the winding roads became gravel, and in the middle of the night sky emerged one of the most beautiful hotels I have ever seen. We were alive and headed to a four-star resort! Phew, what a turn of events. Rich and I gave each other a relieved smile and chuckled.

We checked in and got to sleep immediately, because we knew we had a long day ahead of us. As soon as I awoke, I realized how special this place was. Deep in the mountains, we each had a chic cabin, fireplace, the works. I met Rich for breakfast and we laughed over how dramatic we had been the night before as we enjoyed the best huevos rancheros I have ever eaten.

We finished eating and Niels arrived. Tall with curly hair and a welcoming face, Niels is the world adventurer who happens to be the photographer and editor who would lead us through this journey. He gave us the lowdown, and twenty minutes later we found ourselves riding horses to the top of a mountainous forest where three hundred million monarch butterflies end their annual migration from North America.

It was glorious. Our guide didn't speak English, but I did my best at communicating as we rode. The mountain trail became steeper and the forest grew denser, and suddenly the trees in the distance looked as if they were covered in fur.

"*Por favor desmonte el caballo ahora,*" he said.

We got off the horses. As we approached the furry trees, I realized the trees were not furry at all; they were consumed by monarch butterflies. There was not a square inch of space that wasn't covered by the creatures.

The monarchs not only blanket the trees, but their beautiful wings turn the ground into a fluffy graveyard after the butterflies pass.

"Holy crap, this is amazing!" I shouted to Rich and Niels, who were setting up the camera.

"Shhhhhhhhh!" Our guide hushed me and shot me a dirty look. He must have told me to talk quietly in one of those monologues he'd given that I didn't fully understand on the way up. But okay, will do.

In a whisper, I begged Niels and Rich to get started. I was moved by the beauty of what was in front of me. Whispering, I brought the camera into the scene and spoke about the issues the butterflies are facing. I felt like I was doing a documentary for *National Geographic*, like I was finally doing what I was supposed to do.

All the hard work I had been doing with Dr. Wilson was starting to pay off. Amid these butterflies, I could feel myself fully emerging from the cocoon of depression. I was getting ready to fly.

Later that day, we met our pilots to discuss the paragliding slated for the next day. We enjoyed a delicious dinner, and Niels showed me some of his paragliding photos so I could get a sense of what I would be doing.

The Paragliding World Cup is an annual event where the best pilots in the world gather and take on a challenging virtual speed route in the sky. They carry GPS devices that show them where the virtual buoys are in the air, and they must fly to and around them through a course.

The course was too long for me to fly the entire route, but I was going to get the experience from one of the best pilots in the country. I signed the waiver that reminded me that death was a possible outcome, but it didn't even faze me. I stepped into the bright orange one-piece suit made to keep its wearer warm high in the atmosphere and strapped on the helmet. Then my heart began to pound. Not with nervous energy, but eager energy. *This is what makes me different, and this is what makes life worth living,* I thought. My pilot asked me to stand in front of him as they hooked me into the harness attached to the "wing," the colorful part that looks like a parachute. It was laying on the ground, spread out and ready for our flight. We waited for the perfect breeze to ride up the mountainside and toward us so we could use it to fluff the wing above us. It all happened so quickly. The wing inflated and my guide told me to run.

Off we ran. Seriously, off the mountain.

Step, step, st— I hadn't even taken my third step, and the strong wind had lifted us up at an alarming rate.

It was exhilarating, and once we were up, it was peaceful. And there was another unexpected positive—I was in the atmosphere. It was such a revelation for me. I was understanding the atmosphere like I never had. When I'm storm chasing, I see and feel it all, but only at the surface. Here we were flying on the thermals that I had studied for so long. I could feel the instability helping me fly. We climbed to twelve thousand feet; we were touching the base of the clouds. And again, pure joy and satisfaction rolled over me. This was what I was supposed to be doing. My pilot asked how I was feeling because some people get airsickness, but I felt fine. He asked if I wanted to have some fun. "This isn't the fun part?" I yelled. He immediately collapsed the wing and we plummeted in almost a free fall toward the ground; then he opened the wing and we caught a thermal and rocketed back up. He took me on huge dips and flips, and it felt like the freest roller coaster I had ever been on. We landed without incident, stepping back on terra firma as softly as we had taken off. It was almost too easy. I talked to many of the other pilots that day and took another flight, just for fun, without the cameras. As we made our trip back to New York, I started thinking that I was perhaps supposed to be a paraglider in real life.

This trip was only the beginning. I told Matt when I got

back that I wanted to do more. I helped craft the stories and was so proud watching them play back on TV. This was a part of my job that I hadn't even anticipated but loved so much. From that moment forward, I became the adventure woman along with being a network meteorologist. I went iceboat racing on Lake Minnetonka in Minnesota, NASCAR driving, skydiving, hang gliding, parahawking in Nepal, diving with sharks, crabbing with the guys from *Deadliest Catch*, rappelling down a building, spelunking in the world's largest cave, flying a drone into a volcanic fissure in Iceland, and more.

In each of these adventures, my will to live life grew. And that was saying something. I had come so close, more than once, to taking my own life and not getting to this place. What a shame that would have been for so many reasons. With every flight, every moment taking in some breathtaking vista in one of my adventures, my love and zest for life has flourished.

Of course, the reason Ben Sherwood had hired me was to cover storms. I had been at ABC for almost five months and it was late winter when my first big storm took shape. We were on the cusp of my favorite season to forecast and storm chase: spring, and tornado season in the Great Plains.

I started looking at the computer models in even more depth than I usually do and remember watching as a "huge Friday" started coming together. A "huge Friday" means that even a week out, I could see that the variables that make tornadoes form were already starting to rear their evil heads on the computer models. I talked to Gerard that Sunday and said, "This is the one. I'm going to storm chase." I had learned so much from storm chasing in college and my time with the show *Storm Chasers*, and I was confident I could do it. I went to Tom Cibrowski and asked him to send me to Cincinnati by Thursday for the impending outbreak. Tom is a news lion, aggressive and tenacious; he is a staple of ABC News (he has worked his way up through the company from the very bottom) and just happens to be a fan of all things weather. So, I thought my request was an easy yes.

He thanked me for my excitement and passion but said no. He and the assignment desk already had a plan for me. They really thought it would be best if I covered the snow in Massachusetts.

What?

Granted, it had hardly snowed the winter of 2011–2012, and New England being in a snow drought was a big deal.

But how were three inches in western Massachusetts going to trump multiple tornadoes? Deadly tornadoes!

How could I convince him my instincts were telling me the story would be in the Midwest? I was still so new, so I didn't want to push it. I figured they would see my point in a few days.

But nope, I ended up in Worcester, Massachusetts. Ugh. I kept writing him, though. I said, "Okay, I showed the few inches of snow. Please let me show you what I can do. Send me to Cincinnati tonight for *World News*. I hate to say this, Tom, but tomorrow, people will die." It was a bold statement, but unfortunately most meteorologists knew it was true.

Tom agreed I should go. I flew from Boston to Cincinnati. On *World News Tonight with Diane Sawyer*, I was able to stand in front of the Cincinnati skyline and warn America that a major tornado outbreak was less than twenty-four hours away. I was able to show a concentrated region that needed to be on alert.

Within those twenty-four hours, at least forty-one people were dead, with many more injured. We were there, warning everyone as we chased the storms. We were there chasing the Henryville, Indiana, tornado as it took out Henryville High School. We were there just minutes after the school buses lifted by fierce storm winds skewered buildings. And that night I made it on air with Diane, confirming all that we knew so far. It was my first big storm for ABC.

Later that night I got an e-mail from Diane.

- - - - - Original Message - - - - -

From: Sawyer, Diane
Sent: Friday, March 02, 2012 08:25 Pm
To: Zee, Ginger

Subject: You were dazzling tonight–serious, informative, brave, caring. Bravo.

Diane Sawyer called me "dazzling." And informative, serious, caring, and brave. And she was right. I was finally dazzling and believed it myself. I was healing and becoming the healthy, happy productive woman I knew I could be.

Remember when I told you I make my passwords my goals?

Well, by this point I was fighting that self-destructive behavior with every weapon I had, and positive passwords were a big help. Hey, it had worked for me before with the old *todayshow10* password.

Corny passwords were all fine and good until I had to call my producers back in New York while I was on that tornado chase and ask them to update my password since my e-mail was full.

I was on the phone with Claire Brinberg, a senior producer for *World News*. I was spelling out my password as if she weren't going to figure out what I had just said. . . .

S-P-A-R-K-L-E-E-V-E-R-Y-D-A-Y-12

Sparkle every day. Freaking nerd. But I wasn't a nerd. I was a natural disaster, working through my demons. And sparkling. No, *dazzling*. And that was it. From that e-mail confirmation from Diane until today, it has been a positive ride. Not just six years of luck, but hard work, love of myself and others, a constant desire to learn, and a daily use of my fence. I'm far from perfect, but let me tell you who never picks up a wine bottle before six P.M. This girl. And it is usually just one glass. It's not the answer. I was, all along.

Chapter Eighteen
THE ABCS OF TRAVEL

How many airports have you been in in one day? At ABC it is a crazy game I play with other correspondents and field producers, and we all seem to enjoy telling our travel "war stories." My record is five airports in one day. Let that sink in next time you're standing in the TSA line. I'm not complaining, because I love my job, but it was a day that would have made John Candy and Steve Martin cry.

The five airports that one day were DFW (Dallas/Fort Worth), DAL (Dallas Love Field), BHM (Birmingham), ATL (Atlanta), and JFK (New York). Well before I was a traveling nut for ABC, I'd studied those three-letter identifiers for meteorology. I never would have thought I would have the opportunity to visit so many all at once. On this particular day, we did a live shot in Dallas on floods, then went to DFW to try to make our flight to New York. As soon as we dropped the rental car off, the New York desk called. "Change of plans.

We need you to go to DAL and fly to Birmingham, Alabama, for an interview" (one that I had been trying to get for over a month). No problem.

We landed in Birmingham, then drove an hour to interview a family that had lost their child in a tornado when the entire family had been blown from their home. Heartbreaking. Then we rushed to get the last flight from Atlanta to New York. I was exhausted but exhilarated at the same time. There has always been a calm that I tap into during these endless days driven from a sense of achievement, of doing the impossible. Thanks to my Energizer Bunny of a mom, I never feel like I can accomplish enough in one day. I am a woman who lives with a never-ending list of to-dos and have been told by more than one person in my life that I need to learn to relax. So a day where relaxation isn't possible makes me feel productive and harmonious. To anyone but a natural disaster, this would probably be disturbing. But to me, it's fun, and just another part of my job that I love.

My new role at ABC was full of these productive trips. For example, there's the time I was going to shoot a feature story in Missouri, got delayed overnight in Atlanta, flew to Springfield, Missouri, and drove an hour to Branson to shoot for GMA on a roller coaster. As soon as I got on the roller coaster, I got calls and an e-mail from our senior producer, Chris Vlasto.

"Get off that roller coaster. We need you in Las Vegas by tonight for *World News* and GMA in the morning to talk about the record heat."

I started to explain that I was just getting off the roller coaster, then remembered my New Employee/Good Soldier role and opted for something more efficient.

"I'm on my way."

That adrenaline, the travel test that lay ahead, was unlike any challenge I had been presented in my career before arriving at ABC, and it was thrilling. I loved to make the impossible probable. It was a pop quiz, and I was passing with flying colors.

I used to travel and move without reason to avoid the depression that would settle in; now, as I made inroads with Dr. Wilson each time I was home in New York, the constant crisscrossing of the nation had purpose and never felt like I was running from something, but rather to something. One of the great deterrents to depression is purpose. I was learning that I had it in life despite my career. My identity was growing just as my career was blossoming.

I made my live shot and did GMA the next morning from the Strip, and then an e-mail arrived from Wendy Fisher, the true mastermind behind the positioning of ABC's correspondents. Wendy essentially runs the assignment desk logistics, keeps track of where everyone is and when they are moving, and organizes their travel arrangements. She wanted to know where I could go that would be even hotter. I was so excited to name the city or locale that had been on my bucket list for a long time.

"Death Valley. It could reach one hundred twenty-nine, which is nearing the all-time world-record high of one hundred thirty-four."

"Great, go," she said.

And just like that, my producer Brandon Chase and I hit the road and drove to Death Valley, where the temperature did reach 128°F. As soon as we arrived and started setting up a time lapse of the sunset, we saw a roadrunner and a coyote. We thought, *This has to be fate.* After a 107-degree four A.M. live shot (GMA starts at seven A.M. eastern time, so whenever we go west we have to be ready by three A.M.), we were instructed to go back to Las Vegas. We did another twenty-four hours of storytelling, and we were pretty spent. I always went into these trips so positively, loving every moment of the wildness. But then exhaustion would hit me and I would get a little salty from sweating in these record-high temperatures. Either way, Brandon helped keep my spirits up (he has to have a little natural disaster in him, too, because he has been on the road for at least six years), and we made it through, booked our flights for the next morning, and felt ready for a night of sleep. And perhaps a meal. I definitely get hangry, too.

We pulled into the hotel, and there it was: an e-mail from Vlasto.

Nineteen firefighters were killed in Arizona. We need you to get there by GMA.

Immediately I felt a combination of sorrow for this horrible tragedy and fatigue as I thought about the long drive ahead of us. Again, I was in Good Soldier mode, so all I said was, "We'll be there."

Brandon and I mapped out the drive, which would hopefully be just under four hours. Okay, not bad. But it was already eight P.M. Pacific time, which meant we had just eight hours before GMA started. We hadn't eaten all day and had slept about nine hours total over the past three days. Neither of us thought it was a good idea to drive under the circumstances—exhausted, in the mountains, at night. By this point I had been with ABC for almost eighteen months and was feeling a bit more emboldened. I was more than happy to make the trip, brush my teeth outside the SUV . . . but this was not worth our lives. I called the overnight producers and explained our situation. They agreed it was on the dangerous side and suggested getting us a driver. Then we could sleep and still get to GMA on time. Great idea. But oh, goodness, did we have no idea what lay ahead of us.

The driver arrived and Brandon and I loaded our gear and suitcases into his black SUV. At this point I was so delirious from lack of sleep that it looked like a carriage to my castle. We hopped in, put the seats back, and gave each other a high five. *We can do this,* I thought.

I woke up thirty minutes later and thought it was weird that we were still in a residential area. I looked up front and saw the driver frantically going through a stack of papers. I leaned a little closer to look at the papers and see what he was reading.

MapQuest.

WAIT, MAPQUEST?!

This was 2013. No one used MapQuest anymore. We had smartphones, with GPS. I opened the map on my phone. We were barely ten miles from where we had started. Heading in the wrong direction. I nudged Brandon awake and showed him on my phone how little we'd traveled. Brandon spoke up.

"What route are you planning to take? Ninety-Three looks like it would be a straight shot," Brandon said.

The driver looked back at us, clearly distressed. "Yeah, I just . . . well, I can't find Ninety-Three."

If you've never been to Las Vegas, there is really not much to the highway system. US Route 93 is very easy to find. If you aren't using MapQuest.

We suggested that he use his GPS instead. He said, "I don't have one. I've never really driven anywhere outside the Strip."

This guy had never been off the Las Vegas Strip. In his entire life.

So, instead of sleeping, Brandon and I directed the driver back through Las Vegas to US 93. Once we hit the highway, we both figured, *Oh, good, now we can rest.* There are no turns, just a straight line to Yarnell, Arizona. We repeated the direction, "Just stay on Ninety-Three," and fell asleep.

Two hours later I woke up and it was pitch black. I looked around and saw mountains. That was promising. The driver heard me wake up and said he needed to pull over at the next stop to use the restroom.

Now it was midnight. We had just four hours until GMA started and, I figured, about two more hours to drive. I checked my e-mail and felt even worse than before I'd gotten a little sleep. I opened my map, and lo and behold, we were not on US 93. We were on Interstate 40, going *away* from our destination!

I woke Brandon again. We were both so frustrated, and laughing, because what could we do? As soon as the driver got back in the car, we told him he had messed up again. We got him going the right way, and less than an hour before GMA was to begin, we arrived on the scene where nineteen firefighters had lost their lives.

I flew home from Phoenix the next day after seven airports in five days. As soon as I hit that seat on the way home, I took a breath for the first time in twenty-four hours. I thrived on the pandemonium, yes, but the reality of the stories I was telling was impossible to ignore. I don't know what pre–Dr. Wilson Ginger would have done in these situations, but I can't imagine it would have been as measured. I almost had the best of both worlds now: the ability to live in chaos paired with the tools to separate my feelings when needed.

As a sort of unwritten rule, the first year or two that you work as a network correspondent at ABC seems to be an initiation period of "Let's see how much you can handle." As a viewer, I can always tell when the bosses are liking a new television correspondent, because I see them all the time. The more they like you, the more they have you in action, and that was certainly the case for me when I got to ABC.

For two years, I traveled nonstop, often working fifteen to thirty days in a row, then having just one day off before starting a long stretch again. I can count on one hand the number of true "weekends" (meaning two days off in a row) I had those first two years.

A typical work week would begin with my GMA *Weekend* and *World News* shifts on Saturday and Sunday. Then Sunday night I would fly out to wherever the weather was happening. I traveled so much, I actually had to write down the hotel room number of wherever I was sleeping that night in the notebook on my phone.

One of my favorite travel tales begins with a slew of e-mails from my bosses right after I finished a Sunday GMA show.

The messages instructed me to prepare for a flight to Minneapolis to cover a snowstorm. So, about eight hours later, immediately after I ran off the set of *World News*, I boarded a Sun Country Airlines flight (contrary to popular belief, there are no ABC choppers or private jets; I have always flown commercial) to MSP out of JFK with my work husband, David Meyers. David was a field producer who essentially organized the crew, coordinated the live shots, drove, and was my partner on the road.

David is a solid Midwestern guy with the friendliest disposition. I was fortunate to have him as my travel buddy. We got along so well, and that is important when you spend more time with the people you work with than with anybody else in your life.

After barely making our flight, we landed in Minneapolis and did a live shot for GMA that Monday morning. Then we jumped on another flight to Washington, DC, to cover a significant nor'easter (a nor'easter is a storm where winds blow from northeast to southwest).

When we landed in DC, we did *World News* and GMA, then drove to Front Royal, Virginia, for the epic-snow part of the story. But just as we were going to settle in and get a few hours of sleep, our GMA producers called and said, "We need you to find damage." That meant they wanted a story on the damage from the storm. So we went out into the field but couldn't find anything. The only reported damage was a four-hour drive south, which we relayed to producers along with the fact that we hadn't slept in forty-eight hours. They said they really needed us to get there.

They were looking for the shot of us in front of a fallen tree on a house. We had no idea where to find this, so we just started driving. All night. And finally, like an oasis in the desert, it appeared. We found a fallen tree over a road at the University of Virginia in Charlottesville. Lots of trees, actually. And a tree into a house! It sounds so awful when I think about the reality that our success came in finding another person's property damage, so please understand, I am always cognizant that there are real lives inside that home with a tree on it. We only had two hours until GMA would be live, so we decided to nap in the SUV. I remember brushing my teeth and peeing outside the SUV as if we were camping just before going on air, thinking how glamorous being on network television is!

Moments after I finished the update for daylight (we usually stick around to make sure our shot is updated for the West Coast feed of GMA), we got the call from our producer to go *back* to DC to reposition before the storm hit Boston. In just two and a half days, we had gone to four cities. And by dusk, Boston would make five. As soon as we landed, Chris Vlasto told us to be in nearby Worcester for Thursday morning's GMA. But just before we got to Worcester, we got the call that they had decided against having us on GMA in the morning and that we could turn around and come home. So we drove back to Boston, booking a flight home to NYC as we drove, and just as we were returning the rental car we had taken only ninety minutes before, we got another call. "Sorry. We *do* want you in Boston." No problem. Back we went to Boston in search of a hotel, and then we got the call that they weren't going to use us after all. David and I started laughing at that point. There was really no other response. The upshot? They felt so bad for moving us back and forth that they offered us a night at the Four Seasons. After two nights of restless sleep in an SUV, we said yes pretty quickly. Up until that night, my top-of-the-line work hotel was a Marriott, and most of the time it was more like whatever Econo Lodge came along the way.

After we checked in, knowing we didn't have to work in the morning (another unusual bonus), David and I decided we deserved a nice meal and a few drinks. It felt like the best meal I had had all year, and when we were through I went to bed and slept harder than I had in a year. That is until the hotel phone rang around five A.M.

It was David. The producers changed their minds (again) and wanted us live in Boston in the snowstorm after all. It was snowing heavily at this point, as we had forecast, so driving was nearly impossible. Our crew was coming from Worcester (since that is where we had originally booked them) and pretty soon we realized they weren't going to make it to us in time for the broadcast. So we called our affiliate WCVB and asked them if they had any cameras we could borrow and any lines to connect us to the network. They said yes, and so we drove a treacherous seven miles north to their station.

At this point, I kept pinching myself to wake up, it felt so much like one of my stress dreams. But this was not a dream—this was real life. We got to WCVB and rushed in. They were busy enough with their own coverage. They pointed us toward a locked cabinet. The camera was in there. David had no idea how to use one of those huge news cameras. Plus, they had misunderstood. They did not have a way to broadcast to NYC.

After all that, we were not going to make air.

And then we had a brilliant idea. I genuinely don't remember if it was mine or David's, but I do know that ten minutes later I was live on GMA. Via Skype on an iPad. After the show, I got a lot of great feedback about that iPad broadcast.

There was no time to rest or celebrate. We got straight back into the rental car and drove more than two hours to Cape Cod to get blasted by snow, rain, ice, and waves for our *World News* live shot. Then one more shot for GMA finally concluded our epic trip, which had lasted ninety-six hours and taken us to eight cities. It was stressful, and at times we wondered if we'd pull it off, but we did it. I had to rush back to NYC to be in place for GMA Saturday. And so went my cycle in those early days at ABC.

I always tracked my travels in a journal, and there was one line I wrote on the way back from one of my epic trips: *I just saw a family get on the plane. A mother with her two young children. I wonder, when, if ever, am I going to be able to do that? How? I'm not getting younger, but to keep my job I have to be viable and available always. How does a wife and a mother do that?*

I remember crying as I wrote that line. I had been running nonstop, I was so tired, my body felt fatigued, and I was only in my early thirties. As much as this natural disaster enjoyed the unruly life of a traveling correspondent, there was a large part of me that yearned to settle down. If even for a weekend!

I wish I could go back to that version of me and give her a pep talk. It is going to get better. It always does. All of those things I was so concerned that I would never have a chance to have—a husband, a child, a weekend. I have them now. And I cherish them so much more because I know what it is not to have them.

Chapter Nineteen
EL RENO

I woke up in a panic. After fumbling with my phone in a dark hotel room, my tired eyes finally made out the time—3:45 A.M.

A mental scramble of questions fired through my brain: *Am I late for work? What hotel is this? What city am I in? What time zone am I even in?*

Eventually my brain settled down and I remembered that I was in Wichita, Kansas, and I still had thirty minutes to get up and out the door. I'd been driving all night with my crew and our producer, Gina Sunseri, in search of a live location where damage had been done by one of the tornadoes we had been chasing.

Rolling out of bed, I also remembered that my body was aching from sitting in the car hunched over, writing scripts on my phone, for the last two days. I did some quick stretches before flipping on the light in the bathroom mirror and coming face-to-face with my face. Not good. The lack of sleep and attention to removing my eye makeup the night before had left me looking pretty exhausted. I spackled on some makeup to fake my best at-least-I'm-alive look and rushed out to meet Gina.

After we did GMA from a gas station that had been shredded by a tornado the day before, it was time to focus, because today was the big day. It was May 31, 2013, and Oklahoma City was in the center of the severe-weather threat zone for the second time in ten days.

When we're storm chasing, we forecast for a target region using all the available data on computer models. Then we start narrowing it down by looking at current conditions on satellite. We make our best scientific assessment of the safest place to be closest to the storm, get there, and wait. We watch the tops of the thunderstorms begin to grow, billowing into the sky at such a fast rate, they almost appear like a time-lapse. Once those cumulonimbus storms grow and we can see the base, we match the visual with the radar and find the tornado. Gina had been chasing with me before for *World News*, and we shared a love for the entire adventure—the science, the intuition it takes to be in the right place at the right moment, and then just the awe and respect that is beyond words whenever we get up close to a storm and feel its immense power. The feeling I have when I see a tornado ripping across a barren field is probably similar to the feeling sports fans get when their favorite team wins a world championship. It's exhilarating for me to witness nature's power and watch whatever natural disaster I've been forecasting finally unleash itself. I've had the opportunity to see a few dozen tornadoes at this point, but one of the most stunning images of an up close tornado happened in the boot of Missouri in 2009. This beast was plucking huge trees from the earth like it was a pair of tweezers on a set of eyebrows, tossing the colossal trunks through the sky. Every time I'm out in the field I am reminded of the reasons I made so many difficult choices along the way to avoid being confined to a studio full-time. I know it sounds geeky to say, but I love the weather. I have respect for it, and it inspires me every single time. Today was no different.

Just two weeks earlier, my storm chasing had brought me to Moore, Oklahoma, where twenty-four people had died. I was there just an hour after the tornado hit and the wreckage it left behind was devastating. Moore is famous for the F5 tornado of May 3, 1999. We studied this tornado in college, and it still holds the record of fastest wind speeds ever recorded—301 miles per hour! So when Moore was hit again fourteen years later and we were there, we were all braced for something big. And to be in the general vicinity again on this day felt eerie.

I had a bad feeling that today's tornado in Oklahoma

would also take lives. It's an intuition we all develop over the years by studying storms and the science behind them, and sometimes I wish I didn't have it. It's the type of forecast where I would be happy to be wrong. Unfortunately, that wasn't the case this time.

World News broadcasts at 5:30 P.M. central time, and on this day they wanted me live. Because it takes at least an hour to set up a shot, being live pretty much puts us out of the game as far as getting a tornado on TV or tape. With a chase, you have to move. And most severe storms and tornadoes erupt in the late afternoon and evening, after the peak heating of the day fuels the thunderstorms.

Knowing that I would not be able to chase for a while, I chose a place on the west side of Oklahoma City so at least we could see the storms blowing up while being in the major population center, warning folks of the storms on their doorstep. As we watched the anvils taking over the early evening sky, I had my RadarScope app refreshing constantly. I was watching these incredible images of a supercell (a rotating thunderstorm) morphing into a monster, and again that sense of wonder and respect was familiar yet overwhelming. I knew that the storms were just thirty miles away and they were headed towards us. I also kept refreshing the websites where you can log on and "chase" right along with the storm chasers. You can log on to see their dashboard cameras and watch their vehicles move around the storm on the map. I was so envious of all my friends out there surrounding the developing storm.

As soon as I finished the live shot, I looked at Gina—I wanted to hit the road. We blasted west, but with every refresh of the radar I got more and more concerned. I asked Gina to slow down so I could make a decision as to how to approach . . . or perhaps not approach. The radar was updating slowly, but as soon as the last scan of the atmosphere came in, I shuddered. The structure of the storm now looked like nothing I had ever seen. My stomach dropped. I had a terrible feeling. The freakish blob was centered over El Reno, Oklahoma, and it was so ugly. When I say the radar was ugly, I mean it was far from textbook. I knew how to chase a textbook storm. But this one, I didn't want to mess with anymore. So I told Gina to drive south—away from the storm. I had already directed the crew and satellite truck to go south and get out of this storm's way. Gina and I raced south and within five minutes started seeing the tweets:

The Weather Channel's Mike Bettes flipped vehicle in tornado

Discovery Channel's Reed Timmer's storm-chasing vehicle loses hood

Several chasers getting far too close in the El Reno tornado

I had storm chased with Reed on the show *Storm Chasers*, I knew of Mike, and there were definitely other friends and colleagues I had out in the path of that storm. No one was dead, but it appeared there were injuries.

This had never happened before. Storm chasing had been getting more and more popular, congested, and even dangerous at times. Not because of the tornadoes themselves—but because the chasers were reckless. Yet there was no way all these people were being reckless. I couldn't believe it. As the tweets kept rolling in, my jaw dropped. It was like a nightmare.

My heart collapsed. If *World News* had not insisted on a live shot, my entire crew and I would have been near that tornado, and chances are we would have been hit as well. *World News* might have just saved our lives. So again, thank you, Diane Sawyer. I believe that makes two times you have kept me alive.

Gina and I drove all the way south to Chickasha to allow the storm to pass before taking back roads north to El Reno. Through the flash flooding and postfrontal winds, we finally made it to the town where the storm had hit. It's very rural, which is a great thing, because right away the landscape looked a lot better than what I'd seen ten days earlier when I had arrived right after the tornado in the densely populated town of Moore. Still, there were several deaths and more than a hundred injuries. That evening, I did updates on the storm for *World News* (West Coast), *20/20*, *Nightline*, and *20/20* (West Coast) with all the footage from the damage we found in El Reno. As soon as we finished, I realized GMA was only a few hours away. I couldn't sleep. Not in the van and not after such a wild night. I was too frazzled. Mike Bettes had survived, but as each hour passed, more and more video of storm chasers, people I knew, kept rolling onto my feed. There was wild video of another storm chaser, Brandon Sullivan, barely outrunning the tornado with his group.

All day Saturday we found damage, talked to survivors, and told the stories of yet another deadly twister in central Oklahoma. I was fatigued, but figured I would stay through the weekend and get a flight out Sunday evening. That was before I knew the full truth.

Sunday morning, about thirty hours after the tornado, I woke up to dozens of missed calls from one of the bookers at GMA. All the texts, e-mails, and voice mails were asking the same thing: Do you have Tim Samaras's contact information? Of course I did! I called the booker and shared Tim's info. Before I hung up, I said, "Why do you need to talk to Tim?"

"We are ninety-nine percent sure Tim and his son died in the tornado."

"What?!" Nope, sorry, you are wrong. That's the craziest thing I have ever heard. Maybe it's someone else? No way it's Tim."

I didn't believe my booker for one moment. I called Gerard, the weather producer who had so warmly welcomed me into GMA, who was back in New York getting our show together.

"Gerard, you are not going to believe what the booking department just called and said to me. They tried to tell me that Tim Samaras died in the tornado."

Silence.

"I'm so sorry, Ginger. Tim, Carl, and Paul died in the storm."

Gerard kept talking, but I froze. My world went black with the news that one of my heroes was gone.

Tim Samaras was a highly esteemed and respected storm chaser and scientist who worked on the Discovery Channel show *Storm Chasers*, and that's how I knew him personally. But I knew of him for a long time before that. Tim was a rock star in my world. He was the person I was most excited to meet when I first got to guest on the show. I'd been a fan of Tim's since college and had read his work. I studied his storm chasing the way a musician might study Beethoven or the Rolling Stones. Not only was he a weather genius, but he was the ultimate professional, always conservative and never reckless in his analysis and action the way so many storm chasers can be. His primary focus and reason for storm chasing wasn't the adrenaline rush but the opportunity to collect data that would benefit other scientists. Less than a year before, I'd gone lightning chasing with Tim and his son for five days in a special I was doing for *Nightline*. The trip was so exceptional because we were able to get to know Tim really well. We met him at his home east of Denver, toured his workshop, and saw the giant camera he called the Kahuna—one of the fastest cameras on the planet, designed to capture the birth of the lightning strike, which is still not fully understood in the study of the atmosphere. I ended that beautiful piece for *Nightline* with a quote from Tim: "I don't know how many storms I've seen in my lifetime, but every single one of them still gets me excited. The little boy in me just wants to come out and watch and stare."

I was proud of the piece we did. It was beautifully shot, and it showcased the part of chasing and the science I had always wanted people to see. Tim was the star. And now he was gone.

I hung up the phone on Gerard and began to wail. Gina looked at me and her maternal instincts kicked in. Without asking for details, she immediately held me as I sobbed and rocked in her arms.

We were now a half hour to air and I was still sobbing— the kind of waterworks that are even too much for the movies. I couldn't even catch my breath. I really don't think that I would have experienced more grief in that moment if an actual family member had died. I was way too tired, and this reality was overwhelming and so hard to process.

I knew Tim personally, just enough to feel connected. But it was more than that. I respected him. He was a pillar in the world that I'd devoted my life to, and now he was gone. Up until Tim's passing, I never thought of storm chasing as potentially deadly. From the storms over Lake Michigan that I initially fell in love with to the tornado I had seen just ten days before in Oklahoma, these storms were powerful, yes, but they didn't kill people I knew! No one I knew had died in a tornado, let alone died storm chasing. And for it to be Tim Samaras—the most conservative, non-adrenaline-junkie storm chaser ever? It made no sense, and was incredibly unfair. To this day, I still can't make my peace with Tim's death entirely. It still makes me tear up. Not only did we lose one of the best scientists in our field, but I knew storm chasing had to change. Or at least our relationship to it had to. I had to relinquish control once again to the atmosphere. Mother Nature does indeed know more than we will ever know. That is a very difficult reality for a natural disaster to admit.

I imagined the loss his family would feel and wondered if I had any right to feel my own grief. I was getting closer to going on air, and as much as I wanted to be professional and pull myself together, I couldn't stop crying. I got on the phone with my executive producer, John Ferracane, and we decided that I wouldn't say anything about Tim until his death was confirmed publicly and we were certain that his family had been notified.

Gina gave me a pep talk and I made it through what was the most difficult on-air segment I've ever done, somehow managing not to cry. My voice was quivering, I looked exasperated, but I didn't cry.

For the next twenty-four hours I told stories about Tim, his son Paul, and Carl Young. My colleague, the wildly talented David Muir, who was doing the weekend *World News* at the time, helped me craft the story so Tim, Paul, and Carl were celebrated, not vilified as "reckless chasers." I will forever be grateful for David and his help that weekend. I did my best to fight through the sorrow. I had never told stories about people I knew. And this tornado was so different from any other, and not just for personal reasons. It was the widest tornado in recorded history. It exploded in size during its sixteen-mile track, and most importantly, it reminded us all that no matter how much we learn about the atmosphere, Mother Nature is still in charge.

This was another Katrina moment for me. The humanity of the storm became so much more real, and for me, life changing. This was a natural disaster that brought this natural disaster great perspective and humility.

Chapter Twenty
BEN

One of my favorite adventure stories I've ever done for ABC was the time I went parahawking in Nepal. I flew from JFK to Doha, Qatar, to Kathmandu, and then on to Pokhara, Nepal. The day I arrived I met up with my adventure producer, Niels, and got ready to jump off a cliff, because that's how we roll at ABC. Moments before I ran off the mountain in the foothills of the Himalayas, I had a sudden and unusual pang of . . . well, I wasn't sure what that feeling was. For as long as I could remember, adventure had made so much sense to me. It was like my resting face. So what was this odd sensation that seemed to be telling me that what I was about to do was dangerous? I love statistics, and I had run the stats on this leap. As crazy as it sounds, this jump was less dangerous than taking a cab from the Upper West to the Upper East Side of Manhattan.

I decided to ignore the feeling, and I took those final steps, leaping off the cliff and settling into the seat of a para-glider with a seasoned pilot behind me. I'd done this before at least ten times. I'd flown in the southern Rockies in Mexico and the Andes in Colombia, and now here I was in Nepal, soaring through the sky, the majestic Himalayas my backdrop. I stuck out my hand as instructed, and an Egyptian vulture swooped in from behind us, landed on my gloved hand, and ate the buffalo meat I was holding.

It was a truly magical moment, and as hard as I was trying to enjoy it, my mind and heart were halfway around the world with Ben. Just twenty-four hours ago, my future husband had told me he loved me for the first time. When I say "my future husband," this wasn't the natural disaster talking, jumping off a metaphorical cliff with some guy I'd just met. No, this was my soon-to-be fiancé, my real future husband, Benjamin Aron Colonomos. I love that his middle name is spelled Aron instead of Aaron because his mother, Janis, is an insane Elvis fan. Ben didn't know his middle name was spelled that way, so when he used it as his last name, stage name, he spelled it Aaron. Just before I left for this trip, Ben had said those three words, and I had reciprocated, and now I knew what that pang was. It was love. Real love. Not dramatic, self-involved, dysfunctional love, but a grown-up love. I had found my per-son. I'd done a lot of work to get to a place where not only could I love, but I could be loved. And that's when I realized why I was afraid to jump off that cliff. Because, oh, crap, now someone really did love me and my life really did matter to someone else. There were now stakes for my jumping off that mountain, despite the seemingly low statistical risk. Because now I was jumping for two.

I was at a point in my life where I was valuing my life and my happiness, and Ben had been worth all the work it took to get here. He was perfect timing. Well, almost perfect.

Almost a full year after I moved to New York City, I felt like the best version of myself. I had made some friends, the therapy was working, and I was starting to go on a few blind dates set up by my coworker Dan Harris and his beautiful wife, even though they never worked out. I mean, I love you, Dan and Bianca, but what were you thinking setting up a woman who works on television with a man who takes great pride in not even owning a television? I just don't trust people who say they don't watch TV. Plus, I love Teen Mom 2, which is always on cable. So that would have never worked.

Although I was feeling tremendously better about my life, if you had told me in early August of 2012 at age thirty-one that I was about to meet my husband and in three years would have a baby, I would have laughed in your face. It just didn't seem feasible. I was starting to think I was destined for at least a few years of Carrie Bradshaw-ish aging with grace. Not that

I wasn't worried about watching my eggs mold and decay with no prospect of being fertilized before it was too late. We've all had that image in our minds. Not to mention that as a scientist I'm fully aware of the dramatic scientific charts that show the production and quality of a woman's eggs falling off at an astronomical rate after age thirty-five.

Although that thought and image haunted me as they do many young women in their thirties, I was not on any serious campaign to find the sperm that would fertilize those eggs. I am fortunate to have a mother who, through in-vitro fertilization, became a mother for the third and fourth times at forty-two and forty-nine, respectively.

What happened next was, to say the least, what I least expected.

ABC was coordinating a charity SoulCycle ride to raise money for the American Cancer Society in the name of #TeamRobin. Robin Roberts, our beloved coworker, had been diagnosed with and was undergoing treatment for a type of blood cancer called MDS. My boss, Barbara, sent out an e-mail to invite us all to participate in the spin class.

I think it was a Thursday. I didn't think much of the ride. It was a great workout, and I walked out soaked in sweat and glanced over to see Amy Freeze (the meteorologist at WABC), who lived in my building. I suggested we walk home together. She agreed and said she had a friend with her. And there he was. Amy's friend. He was tall, with the fullest head of dark, beautiful hair, a handsome face, and the most joyful smile.

"I'm Ben. Nice to meet you."

"I'm Ginger. Nice to meet you, too."

It was that simple. From that moment on, as we headed south on the Upper East Side to cross Central Park, we completely forgot Amy was with us. (Sorry, Amy.) Ben and I talked and walked home like the oldest of friends. *This guy is perfect,* I thought as we made our way through the winding paths of Central Park. In no way were we taking the most efficient way home (now I know that's because Ben is horrible with directions, but I also like to think it was because he wanted to spend more time with me), but I didn't care. I wanted this walk to last forever. He explained that he worked at NBC as a feature reporter. He was from a suburb of NYC but had moved around a lot for radio and TV jobs. I immediately started sharing my weirdness with him and felt so comfortable being in his company.

His wit and dry humor were comforting, and I laughed harder than I had in a year. Just as I was finally feeling like myself, suddenly here was this guy who felt like he could make me even better. *Uh-oh. No, Ginger. Stop. Don't jump. No more natural-disaster Ginger. This is grown-up Ginger. You know better.*

But still, as soon as I got home I went on a Google tirade consuming everything Ben on the Web. He was *hilarious.*

He is undoubtedly the most talented person that I have ever known. Ben does all of his segments alone. Okay, he has a photographer, obviously, but he produces every piece by himself. He starts at 7:30 A.M. with an idea, writes, shoots, and then edits it himself to air by 12:30. He has a show that airs on Saturday evenings in New York City called *Life According to Ben*. As Ben explains, it comes on whenever they run out of Shake Weight commercials, and it is hysterical. I could tell he was insane, in a perfect way. I needed more.

I immediately wrote Amy to ask for his e-mail, and I told him in no uncertain terms we should be friends. Yes, I used the word *friends*. After all, I was now grown-up Ginger, and I didn't want to jump too quickly. I didn't even know for sure that he was single. Or straight. This was NYC, and I had been burned by that before. I was not going to assume anything, but I was also not going to go one more hour in my life without knowing this man better in some capacity.

I'll let Ben give you his reaction to my first e-mail.

You know all those times you meet someone and hope that somehow they will miraculously contact you although you didn't exchange information, yet it never happens? Well, it happened. She wrote. And it was amazing. She had me before the letter, during the letter, and at the end of the letter when she said, "We should be friends." A statement I quickly responded to in my mind with "Not a chance." She was the one. The question now was, how do I turn this potential "pal" into my girlfriend?

A few days later, we met for lunch (the "friend" meal) at Ed's Chowder House, which is a restaurant on campus near ABC. I wore my best J.Crew outfit: orange pants and a striped blue shirt. We ate lunch, had a drink, and laughed a lot. He was definitely single. It felt like a date, and unlike the blind dates I had been on earlier that summer, this felt like a date I wanted to go on forever. We walked to Magnolia Bakery and got banana pudding. I had walked past this famous spot so many times but had never given into the temptation. But Ben is the kind of guy who eats banana pudding in the middle of the day, and over time he's helped me relax and enjoy the small but awesome pleasures of life—like banana pudding. We took our pudding to the fountain at Lincoln Center, sat on the bench that surrounds the iconic fountain, and just took it all in. Ben put one of his earbuds in each of our ears, and we chose people who fit the music that came on. We were rolling with laughter.

And then he looked at me, and I knew I was already falling in love. But it felt different from anything else I had ever felt. It was measured, settled, and real. I could already see in his eyes that I was his and he loved me. I knew he was going to embrace all of me, even the natural-disaster parts. It was like we had skipped ahead twenty years. I could see the camera pulling up above us over the entire gorgeous city as our "falling in love" soundtrack played in the background.

And then he did it. He leaned in for the kiss. And it was . . . terrible. Don't worry, not sloppy, not gross . . . but like a kiss I would have given my boss on his cheek. Nothing more than a peck. This is how Ben's remembers it.

> When a girl says that magically platonic phrase, "We should be friends," and you know you want something more, it's important to make a move but not swallow her face. I was testing the waters. It was a peck, then within half a second, once I realized she wasn't wincing in agony, I kissed her again.

The second kiss was for real, as I kissed him back. It was magical, and every kiss since has been enchanting. Truly. I am transported when we kiss, because it feels like home. That's the only way to describe how I feel when I'm with Ben. Home.

After that date, we went on a few more. And then natural-disaster Ginger got involved. Ben was so serious, so into me. It was frightening. There were zero games, I could feel his love, and he was a genuinely wonderful guy. So I did the only thing that made sense.

I broke up with him.

He was home, and remember, as Dr. Wilson had helped

me figure out, I wasn't the biggest fan of comfort and being "at home." I needed more time. More time to make sure I didn't want to find another jerk to treat me poorly for a bit longer. There were plenty of them out there, I knew it! I could give them someone to beat up for a while. I didn't deserve this perfect man. He couldn't possibly be real anyway. The other shoe was bound to drop sooner or later. So I ran. And I regretted it immediately. I was worried I was going backward after all my recent progress, and tried my best to work it through with Dr. Wilson.

About six weeks later, my dad and brother were visiting, and one night in a restaurant, we saw Ben's show playing on the television. I took a calculated leap and texted him to see how he was, and I was thrilled that he was open to seeing me.

We decided to start over and make this one our first date. Ben rented a car and surprised me by picking me up in front of my building. Now, for those of you who don't live in New York City, this is a huge deal. I had given up my car more than a year before this, and I often felt locked in to Manhattan. Hopping in a car was like riding a really cool magic carpet that could take me to so many places I had never been. Ben drove us to Piermont, New York, where he had grown up. We ate at the restaurant he had worked at in high school and college. He pointed out all his old haunts, his school, the corner where he had his first cigarette and first kiss (not on the same day). And then we passed his childhood home. There was a woman standing outside, and she invited us inside when Ben told her that he'd grown up there. To this day, I still give him a hard time about taking me here. I mean, if touring your childhood home doesn't scream twentieth date rather than second "first" date, I don't know what does! But Ben is Ben and he doesn't apologize for loving or feeling, which is just another amazing quality of his that I admire. It was a great day, and when we got back to the city, we dated for a few weeks, until I broke up with him again.

Ben loves telling people I broke up with him twice. And I love my perfect comeback. "Yes, I did break up with you twice. But I then married you and had your baby." Boom.

The second time we got back together, Ben took the lead. It had been almost three months since I had last talked to him, and I had changed my hair color. It was long and very dark now. Ben saw me on television and texted me to let me know how much he liked it.

I was still in therapy, and as much as I thought I'd grown by the time I first met Ben, losing him had set me on a path to do even more work with Dr. Wilson. I felt so much more ready to be loved and love in return than I had the day we met. I texted him back, and whatever I said, I think he felt like I wasn't going to run away this time.

Ben is the type of guy who makes sure you are okay before he is. He regularly rubs my head or feet, always knowing what

I need even before I know it. He has taught me to continue to grow as a communicator. He isn't perfect by any means, but he is still the same wonderful person I met that first day. He is insanely talented, handsome, genuinely sweet, and joyful. My favorite part of Ben is his ability to be independent. He can take care of himself and he never begrudges me the demands of my career. In our first apartment together, he would tuck me in and say good night, then close the door. I would just be getting the comforter up to my chin when I would often hear the most beautiful sound. Laughter. Real laughter. Billowing out of Ben. He was probably laughing at one of the stupid movies he watches over and over, like *The Benchwarmers*, but it didn't matter what made him laugh; what mattered was his ability to laugh and feel joy. So few people can let themselves go and do that.

That joy now gurgles from my son when he is laughing at us. It is the joy that I can now allow myself to have because of all the work I've done and the time I've spent with Ben, and I'm so grateful for that.

Ben and I have the most open, trusting relationship I have ever had. We tell each other pretty much everything but don't overcommunicate. There are many days where we don't even talk until we see each other in the afternoon. Because we don't have to be in constant contact. That is refreshing, and I realize that is a trust we didn't even have to discuss.

By no means do I consider myself a relationship expert

(let's be honest, this is one of the only functional relationships I have had in my life), but I do believe Ben taught me the most crucial variable needed for a successful relationship, and that is respect. First, respect yourself—don't ever let anyone treat you as less than you deserve. Second, respect them. I respect Ben more than anyone I have ever been with. He deserves my utmost respect and love, and I am honored to deliver that every day.

Chapter Twenty-One
KEY WEST

As much as I loved my new job at ABC, my schedule was exhausting. For eighteen months, I'd been working three weeks straight at a time with no more than thirty-six hours off. I was really looking forward to five days in Key West with my new boyfriend, Ben, and my best friend, Brad. That's right. I did what seemed like the most normal thing in the world to a natural disaster, and I invited my gay ex-boyfriend/current best friend to join me and my future husband on our vacation. Although at the time I thought all I wanted was for the two people I cared about the most in the world to meet each other, I have come to appreciate that perhaps the rest of the world would judge my unique entourage as the stuff of a wacky (okay, dysfunctional) romantic comedy.

Ben and I had been dating for six months (after the two breakups) and he seemed too good to be true. Ben was turning out to be the guy the girl meets in the first act of the movie whom you know the girl should wind up with, but she has to spend the rest of the movie realizing the broken, non-committal guy with the constant wink is no good for her. Ben is stable and loving, with a close-knit family who share each other's Facebook posts. They took me in as their own almost as soon as Ben and I started dating. But most importantly, Ben never for one moment kept me guessing how he felt about me.

Brad, on the other hand, had gone back to being my best friend. Even as Brad and I had moved to different cities during the course of our careers, and even taken three years off from communicating at all, somehow we had grown even closer. We are both ambitious and competitive, and share the same twisted sense of humor. Brad was finally back to being my three A.M. friend—the person you can call when your latest relationship blows up and pieces of you are all over the floor. So inviting Brad to Key West with me and Ben seemed like a great way to celebrate my finally getting it right. In my head, it was a long montage of margaritas by the ocean, starring Will and Grace and Harry Connick Jr.

Brad, however, did not see my motivation for the trip as simply as I did.

She wanted me to sign off on Ben. I knew I had the power to make or break this relationship. It was a lot of responsibility and I had to keep it all to myself.

For the record, Brad does not keep anything to himself. Especially after a few drinks. Brad is boisterous and opinionated and, let's face it, a little gossipy. Even before he takes the first sip of a Macallan neat or a Ketel One on the rocks. By drink three, he's a gay Edward Snowden. I love Brad, and even though he can be a bit of a liability, it's a price I happily pay for our friendship. In any case, I deliberately did not want to ask Brad for his advice about Ben. I wanted to keep this trip light. All I really wanted was for Brad to know I was having a grown-up relationship and handling it all like a grown-up. At least that's what I told myself at the time.

Looking back now, I did want Brad's approval. He wasn't just my best friend; he was a bit of a father figure, too. Because I trust his judgment so much more than my own, I wanted confirmation from Brad that Ben was as good as I thought he was.

I wasn't the least bit surprised that they hit it off right away. Brad's joke about who would be sharing the master bedroom in the suite made Ben laugh out loud. Of course they liked each other. How could two people I liked so much *not* like each other? That's just math. The three of us checked into the suite, and after Brad put away enough clothes for a summer jaunt through Europe, we headed down to the bar. Our vacation was off to a great start, and everybody seemed as happy as a basket of puppies. We hit all the vacation markers—drinking, eating, making fun of the tourists in sandals with socks on. I'd love to say there was a scene of the three of us holding hands and snorkeling, or sipping from the same giant margarita glass, but it really just felt like three people in tropical weather chilling out and having a great time. And then, on the third day, when Ben left the pool to take a call, I did the one thing I had promised myself I wouldn't do (because I wanted to play it cool). To be fair, I was at the peak of my vacation good vibes and the bottom of the margarita pitcher when I leaned in to Brad and popped the question.

"So . . . what do you think?"

Brad looked at me like I was speaking Japanese. So I added a few words for clarification.

"What do you think about Ben?"

"Oh . . . Ben!? I thought you were talking about the towel guy. His name is Carlos, by the way."

"Hysterical. He's great, right? I knew you two would be friends."

Brad adjusted his lounge chair to an upright position before launching into what I have to assume was a well-rehearsed monologue. "Listen, Ginger. I didn't accept your frequent-flier miles and the second bedroom in your suite to make a new friend. I'm here to work. For you. The good news is, I have already reached my decision. Yes. My decision is yes. Date him, Ginger. Don't fuck it up, Ginger. Yes, he's suspiciously good-natured, easygoing, and grown-up. He is eerily like me, but really straight. And while I'm not at all sure this will work for you in the long run, because we both know that you function best with disaster and chaos and destruction, if you can somehow fake being a grown-up enough to handle a great man who loves you, you can and should do this. Plus, your eggs aren't getting fresher, and you've never dated a guy with a real job who is a fan of yours both professionally and romantically, so I say why not give it a try, at least to check that off your bucket list?"

He smiled, pleased with his monologue and performance, and reclined the chair before closing his eyes. I didn't say a word, even though I wanted to tell him he was way overestimating his power over me and way underestimating my ability to be with a guy like Ben. But instead I just laughed. Loudly. I put my whole body into it, knocking a perfectly good bowl of chips and guacamole onto the floor.

And then I saw Ben heading over to my lounge chair with a hat he'd bought me from the gift shop. It was a really cute floppy hat that was both fashionable and functional. This was so my Ben. In my whole life, I'd never had a guy who would even notice if I was sunburned, much less care enough to do something about it.

That's when Brad took the hat and put it on.

"How did you know I freckle in the sun? You are the best!" he said.

Brad looked at me and I knew what he was really saying was *He really is great, Ginger.* I got it. And he was right; it did feel pretty good to have someone genuinely care about you.

Later that afternoon, Ben started packing because he had to get back to work in New York. I still had planned two and a half days more with Brad in Key West. I'd realized that the real reason I'd brought Brad was to assure me that I deserved Ben and wouldn't mess it up.

But first I had to say goodbye to Ben.

"So you're totally cool with Brad and me staying here for an extra few days?" I asked.

"Why wouldn't I be? Look, if I have to worry about another man, even a gay ex-boyfriend man, and my girlfriend hooking up, then there's something fundamentally broken with our relationship that cannot be fixed. And then at least we know that and can move on with our lives," he said.

"Which would be terrible," I said.

Ben brightened like a kid who'd just gotten a pony for Christmas.

"Because I love you," I told him.

Okay, now he brightened like he'd gotten two ponies.

I hugged him as he got in the cab to the airport and met Brad at the pool. He was with Carlos, discussing how much Downy the hotel used in their towels to get them so soft, and I jumped right in and played wingman. The three of us had dinner that night, and later Brad and I went up to the room and watched the Kardashians till three A.M.

I was wearing the hat the whole time.

"Not a lot of guys would leave you in a hotel suite with your hot ex-boyfriend," Brad said.

"That's because Ben is a grown-up, Brad. And this is a grown-ass woman you're looking at. Also, you are really gay," I told him.

Brad nodded. If he were a peacock, he would have fanned out his feathers and knocked me off the bed.

"You're welcome."

Brad and I had a great time the next two days. We both knew this might be the last time we would be on spring break together. Grown-up life was taking over, and I was going to be with Ben now. We were probably getting married. Brad cried at the airport when we said goodbye. Because that's how all romantic comedies end. At the airport.

Six months later, Ben and I got engaged. And no, we did not take Brad along on our honeymoon.

Chapter Twenty-Two
GRATEFUL

The stage manager gave me a thirty-second warning and I lowered my head. At ten seconds, the audience detonated into a deafening applause. I heard the *click, click, click,* and then the announcer: "Ladies and gentleman, *Dancing with the Stars* continues our legendary talent contest with the delectable, the defiant . . . Ginger and Val!"

I looked up and started dancing on a stage in front of millions of people. Head to toe in obnoxiously awesome turquoise sequins, I was jiving in front of America. About twenty seconds in, I blacked out, and when I came to, I was in my final dance pose in Val Chmerkovskiy's arms, feeling about as high as one can get without drugs. I was on season twenty-three of *Dancing with the Stars*, and although I had never done anything like this, I felt oddly at home.

My journey to being a contestant on one of the longest-running reality competitions had begun two years prior. It had been three months since I took over for Sam Champion on weekday GMA when I joined our post-Oscars GMA extravaganza live from Los Angeles. We were going to have some of the dancers from *Dancing with the Stars*, including Derek Hough, do a big opening number. One of our senior producers, Margo Baumgart, thought it would be fun if I danced with Derek. I was a fan of Derek's and the show's, and it seemed like a great way to display another side of me to the GMA audience. There was only one problem: I had never done ballroom dance before. But like everything I had said yes to in my life, I threw myself into this challenge with an enthusiastic "Absolutely!"

Derek and I danced, and it was magical. He taught me a one-minute number and essentially puppeteered me across the stage, pushing and pulling me in my gold-sequined fringe dress. I guess I did okay, because afterward, my agent and the producers began talking about the possibility of my being a contestant. It was decided that I was going to be on season twenty. There were rumors that Derek was going to be my partner, and I was stoked. This was going to take my career to another level and allow people to get to know me beyond the weather.

Two days before I was supposed to sign the paperwork for the show, a bombshell dropped. Due to some scheduling issues and other concerns, my bosses thought this was no longer a good time for me to do the show. I was furious, and I didn't understand. I know now when they say, we were not trying to hurt you, and there may come a time that it makes sense again in the future . . . that they were right.

Dancing or not, I stayed plenty busy. One month later, I started hosting a new series on GMA called "Extreme Zee." I would present a series of extreme stunts and travels for GMA, starting with dangling off the side of a Chicago skyscraper and rappelling down the side. I would go on to dive with forty sharks, bungee jump, zip-line off a platform hundreds of feet above an NFL football stadium, and fly a drone into a fissure in an Icelandic volcano.

In late winter of 2015, the GMA producers decided that they wanted me to travel to Jellyfish Lake in Palau, and I jumped at the chance. Palau is a chain of more than three hundred islands between the Philippines and Guam. It takes more than thirty hours to get there, and I would only have six days total to travel and do the piece. As detailed earlier, I had been through the travel wringer at ABC, so this was nothing, and I couldn't wait.

As soon as I arrived, I was shuttled to our hotel. The salty ocean air rushed over me and almost made up for the brutal jet lag I was suffering from.

I immediately met my producer, Jennifer, and we boarded the boat that took us around the stunning landscape on and under the sea. Each of the sandstone islands juts up out of pristine turquoise water, covered in vibrant green mangroves and local plants, including lush flowers that I had never seen before.

We went scuba diving in what divers around the world refer to as one of the Seven Wonders of the Underwater World. On our dive we saw sharks; a plethora of fish; and perhaps my favorite animal ever, a giant aqua-colored Napoleon wrasse. It is a mammoth fish, weighing four hundred pounds, that has a face like an old man. This one followed us through the entire dive, even when our guide brought me to the edge of what looked like the Grand Canyon underwater. As I held on to a chain that was attached to the edge of the rock we were on, they warned that I needed to stay on this leash or I could get swept away in powerful ocean current. It felt like I was holding on to a pole in a hurricane, and I knew it without even seeing it: the video of this expedition was fantastic.

The second day was spent on boats for at least ten hours. We snorkeled and showcased the beauty of Palau. As I was taking my final boat to Jellyfish Lake, I started to feel a little sick. I shook it off and made some excuses. Maybe it was the jet lag, the scuba nitrogen/oxygen, the tilefish, or the wine I had had the night before. Whatever the case, I needed to fight through it, because we were now at the jewel of Palau, Jellyfish Lake. From the surface, the lake appears to be like any other, but as soon as I dove in, I saw them. At first it was just a few orange globes, but as we swam toward the middle, a few became hundreds, then thousands of nonstinging jellyfish. They swarmed around me, and it felt how I imagine taking a bath full of Jell-O would.

Jellyfish Lake was created at the end of the last ice age when a certain breed of jellyfish was trapped in this mostly saltwater lake. Over generations, thanks to a lack of predators, they evolved away from having their stinging cells. It was sublime, and it almost made me forget about how sick I was feeling.

As soon as we docked, I headed back to the airport to catch a flight back to Guam and then Tokyo. During a seven-hour layover in Tokyo, I started feeling really sick and was worried it was food poisoning. And then it hit me. I fumbled for my phone. Oh, my goodness, I was late. Not late for my flight . . . but *late*. I was pregnant. I searched around the Tokyo airport for a pregnancy test but couldn't find anything. I waited the full seven hours in the airport, plus the thirteen-hour flight home, plus two hours in traffic from JFK, to run to the pharmacy and confirm what I already knew. Yes, I was pregnant! I couldn't wait to tell Ben, but decided that just announcing it to him like a lunatic when he got home wasn't good enough. Instead, I made Ben a video using the GoPro footage of my dive in Jellyfish Lake.

As thousands of jellyfish cloud the picture, I emerge in my snorkel gear. And suddenly words start scrolling from the bottom of the screen that read . . . *Even though I wish I could have been with you, isn't it great that your baby was in Palau? Not me . . . our baby.* And then the test popped up on-screen reading POSITIVE.

Ben was stunned by the video. He cried at the big news, and just like that, from that moment forward we were parents.

It wasn't too much later that Ben decided he didn't want me to go on my next ABC adventure in Vietnam. I was going to hike into Hang Son Doong, the largest cave in the world, and do a live broadcast. This is a cave that had been discovered only a few years before and was so dangerous that only three hundred people had ever been inside. It would be my most intense, and possibly most dangerous, adventure yet.

I tried explaining to Ben that the trip had been set for months. I couldn't not go. Ben didn't understand that. All he saw were his most precious assets, me and the baby, deliberately throwing themselves into danger. Also, because I was pregnant, I couldn't get my immunizations or travel shots, which meant I wasn't going to be able to eat any of the fresh food. I was supposed to leave in just ten days and go across the world, and Ben was terrified. I assured him I would never take any risk that I felt would endanger our child, and he reluctantly helped me pack. I told the producer in charge of the segment about my pregnancy, and she helped make sure I would have enough packaged food to make it through the trip.

By the end of our trip, I was sure I could never eat another can of soup or granola bar in my life.

I flew through Seoul, Korea, to Hanoi, Vietnam, and took another flight to Dong Hoi, where I met up with one of our most talented field producers, Bartley Price. Bart has covered everything from war in the Middle East to tornadoes with me, so Vietnam for him was just another stop on the map.

After spending the first night in Dong Hoi, we drove two hours to a small village where we were prepped for the expedition. They gave us a choice of hiking two days through the jungle, or choppering in on a huge helicopter. We chose the latter, as we knew we had a huge hike ahead of us once we got underground and wanted to reserve our energy. This helicopter had red, white, and blue paint, but when I took my seat and buckled up, a piece of the white interior chipped away. It was military green, and I realized it was really that old—you know, from the Vietnam War. The people were so kind, but with my new cargo in my uterus, I was extra cautious and filled with questions and concerns. I watched the dense jungle over rolling hills become more and more rural, and I breathed a sigh of relief as we landed safely a few miles outside the cave. We hiked through and along a river for an hour in the intense heat, and I felt okay. They warned us of bugs, tigers, monkeys . . . you name it. Within an hour, we had made it to the mouth of a giant cave shaped like a crescent moon. I said out loud, "Well, this isn't so bad." And then our guide, Howard Limbert, who was one of the first to discover Son Doong, laughed.

"Oh, no, dear, this isn't the cave. We have days of hiking ahead of us."

Great. We crossed under the arch, and the temperature immediately dropped from near ninety to seventy. It almost felt cold. We walked to the edge of the first drop, and hundreds of yards straight down, you could see the first base camp. We hiked down the slippery rocks, my blood pressure out of control with all the new blood for the baby in my body. My feet hit the white sand beach at the bottom, and I was so happy I had come. This is the part of my job I cherish—seeing parts of the world I would never see on my own and sharing the experience with an audience that may never get this chance. I mean, a white-sand beach inside a cave?! This was already one of the wildest places I'd ever seen, and we were only in the entrance. The guides had warned us of foot rot, so we made sure to dry our shoes and socks as much as possible that night. They warned us that the water at the base camp here was the last we would see for four days. So if we wanted to freshen up, now was the time. As I made my way into that chilly water, I was instantly transported back to Lake Michigan, where my dad used to force me to dunk myself into the sixty-degree lake to wash my hair when we were camping.

That night, as everyone ate amazing Vietnamese food cooked by our guides, I poured Aquafina into a packet of dry soup and prayed that my fetus was getting the nutrients he needed.

Just as I went to take my second bite of the barely cooked soup, *splat*, a swallow pooped on my spoon. It wasn't bats in this cave that tortured us, but a fast-flying bird that followed us and pooped on everything for the next seven days. It was funny and charming for about thirty minutes, and then everything we owned was covered in guano.

That night, the sound of the cave was deafening. There were swallows pooping everywhere and strange animals creeping and crawling around us, and after one forced hour of sleep (I was still on NYC time), I woke up to the fullest bladder I've ever felt. For any of you who have been pregnant, weren't you surprised by how early you had the constant peeing sensation? Well, mine came to me in a tent on a borrowed one-inch-thick mat in a sleeping bag. I exited the tent with a headlamp as instructed and was rapidly swarmed by a dozen species of insects as I awkwardly walked to the compost toilet.

That's right, I said *compost toilet*. We had just entered a national treasure, so of course we couldn't just go anywhere. Instead, we shared a bucket full of rice husks that absorbed our waste. As I approached the makeshift tent over the bucket, I quickly pulled down my pants, and as I started to pee, a creature more frightening than any I have ever encountered fell from the top of the tent right between my legs. It scurried back and forth, hitting my feet, the bucket, and the edge of the tent. I muffled my scream, knowing I didn't want to wake the thirty others from my crew and the guides who had set up camp with us. It was the biggest centipede I've ever seen. This thing had to have been a foot long and seven inches wide, an ugly earthling that was my first real indication that we were not in New York, or even North America, anymore. It was also the end of my using the compost toilet at night. I started using one of my food containers as my very own private toilet inside my tent. I would empty that small container into the compost toilet every morning. The combination of the smell of my own urine I was carrying back and forth with the smells that surrounded that compost toilet was just about the worst thing you can do to a nine-weeks-pregnant lady.

Yuck.

Thank goodness we moved on that day and I learned to strap a scented baby wipe under my nose. Despite the scent challenges, the hike was transformational; each step I took opened a new world that so few had ever seen. And you could feel it. New species were constantly being discovered under this giant crevasse. Physically, it was treacherous, and we had to watch every step very carefully. I was probably overly cautious, given I was pregnant and nobody else on the trip had any idea.

We made it to almost the end of Son Doong and shot some of the most epic scenes for my story. At one point, I climbed to the top of a stalagmite covered in bright green moss. With the drone camera circling around me, capturing the single beam of sunlight pouring into the cave from a hole above, it looks like a scene from *Avatar*. Each scene we shot was more unbelievable than the last. As we made our way back to the cave entrance, I knew we had great footage that would complement our historic live shot to come (few had braved Son Doong, and no one had taken on the task of bringing it live to the world on a broadcast). By this point it had been almost four days since we had bathed or seen the sky. We set up at the original base camp, and this time I dove into that pond with zero reservation, embracing the chill. All the video we had shot so far was transferred back to NYC using a complex series of satellites reflecting off each other up and out of the cave. Our live broadcast was thirty-six hours away. We tested several different locations for live shots and ran through rehearsals with the two awesome drone operators who had captured the aerial journey and were helping us make the live portion look even bigger and more marvelous.

After the live broadcast, I felt that same pang I had years earlier when I realized how much I loved Ben. This time, I was homesick and knew that my life was evolving, and as much as

I love adventure, it was time to get home and grow this little miracle.

By the fall of 2015, as I was nearing my due date, *Dancing with the Stars* approached me about going on the show that spring. My bosses agreed this time, and now all I had to do was get it through my head that I was due to give birth around Christmas and that by mid-February I would start filming. Sure, sounds like a perfectly normal turn for a postpartum woman. So I said yes, because that is what I do.

My son was breech and wouldn't budge, so we scheduled a C-section. Anyone who knows me knows that I can ride a helicopter into a jungle in Vietnam, yet an IV at a hospital is my biggest fear. I had to get one, and once that was complete, the rest was seamless. Except the part where they were already cutting into my belly and I could feel everything. (For those of you who don't know, think of how they numb your mouth for dental work, but you still feel the pressure.) And Ben was still not in the surgery room. I was crying and asking where my husband was and kept telling myself that I needed to stay positive so my baby boy could come into this world in a light-filled room. Ben finally arrived, and when he saw that I had been crying, he made me laugh by telling me that all my ex-boyfriends were in the waiting area.

Moments later, I felt the doctor and nurse wiggle and jiggle my insides, and then I felt him get pulled out. I was overjoyed by the first sounds of my baby boy crying. He was finally here! They held Adrian over the sheet for two seconds like in *The Lion King* before whisking him away to clean him up. When they brought him up to my face and put his cheek against mine for the first time, it was the most instant love I have ever felt. It haunts me how soft and perfect that little cheek was. I think about it all the time. That moment that I became a mother was the best in my life. And even now, as he is reaching for this keyboard and screaming at me for not allowing him to ruin this text, I love every screech. I cherish every tear, every laugh, and every smile. He is a constant reminder that all the sorrow, hardship, and mentally broken moments I have had are nothing. He is my purpose. Through all the relationships, the clawing my way through TV stations, this was where I was meant to be, and these were the people I was meant to be with. As soon as I touched that little face, it was gratitude that I felt. And that is really what life's about. I just didn't know how to express that until later that spring.

So we went home and became parents. Six weeks later, my doctor cleared me to walk around and exercise normally (I still had quite the scar). Two days after that appointment, I flew to Los Angeles and met my partner for *DWTS*, Valentin Chmerkovskiy.

The producers did a great job at keeping my partner a secret from me. I was still carrying a considerable amount of pregnancy weight and hardly felt like myself, but it's amazing what becoming a mother does to your self-esteem. It makes you feel like a superhuman. I felt like I could do anything and everything, even more than I had in my past. So when I walked in and saw Val standing behind the curtain, I nearly fell to my knees. He was the last person I thought they would give me for a partner. He usually got the tough, cool chick with attitude. I was the furthest from that. I ran to him, breathing in that sweet cologne and *eau de Val* that always surrounds him, and gave him the biggest hug. (I was sweating like a little piggy, because I had on a leather coat over a dress over some major Spanx.)

Starting that day, we rehearsed four to seven hours every single day. The physical work for *DWTS* is intense, but the mental work was even tougher. With the new baby at home, I went back to work one week before the premiere of *DWTS*. The plan was for me to fly to Los Angeles late on Saturdays so I could be with the baby as much as possible. Then I'd do camera blocking and prep on Sunday, do the show Monday, and fly back Tuesday morning to make it back for *GMA* and *World News* Wednesday through Friday. The schedule was intense. I would do *GMA*, rehearse with Val, do *World News*, get a few hours of sleep, and do it all over again. Any free moment I had was spent with Adrian and Ben, and as wild as it sounds, I feel like this unchained schedule was great for me.

I was forced to do something I had never been able to do in my life: live in the moment. I became extremely focused and present in my own life. When I was with the baby, I never picked up the phone to text or e-mail. When I was with Val, I was taking in every ounce of instruction and advice he had for me. This focus spilled over to my husband and my "day job" as well, and as crazy as that period of life was, it was equally as exhilarating. I have learned in life that being busy is all relative. For me, it's just what you bring to your schedule, your choices, and your life that makes the difference.

Val is a great friend to this day. What most people see on TV is a hot guy with his shirt off, but I am here to tell you he is ferociously talented well beyond his dancing ability. He is a poet, a rapper, a violinist, and a creative force who loves inspiring young people and giving back to the community. As awesome as he is, he has a distinct style of teaching that I am not sure I could have handled at any other point in my life.

Val and I were always buddies off the dance floor, but the moment I stepped on that floor I had to be ready to work. I am obsessed with commitment, so our dynamic was harmonious. If you are lucky enough to get Val as your partner, you better know how to take criticism and not break under pressure. Using the tools Dr. Wilson taught me, I would block any emotion I thought he felt and stay in my own lane doing the best I possibly could. At a certain point, Val asked why I didn't react when he was coaching strictly. I told him about the relationships I had been in, the tools I had learned, and the beauty of timing. Had I done this show five years before, I would not have been able to handle Val. First of all, I probably would have flirted with him. Second, I would have broken down into a puddly mess of low self-esteem any time he critiqued me.

Whenever Val and I were about to step out on the dance floor, he would take my hand and say, "When are you ever going to get to do this? Look at this ballroom. Hundreds of people will be filling this room tonight to watch you. Millions will be watching at home. So few people get the chance to do this—don't take it for granted and don't forget to be grateful. Dance with gratitude."

Val did this on our first show and I had an epiphany. Gratitude was the simple word for what had changed in my life. It's the feeling I got when Adrian was born. The feeling I have for Dickhead despite how he hurt us; the feeling I have for my fruit-chucking mom; the feeling I have for Valerie, Sheree, and all the mean viewers over the years. The overwhelming feeling I have for my husband, Ben. But most importantly, I am grateful for myself, and I am proud of all that I have gone through and of the woman I have become. I am grateful for life and want to pass it on.

And that's the point, I guess. Being grateful for it all, good and bad, because it got me to where I am now. It's not the annoying hashtag #blessed, either. It's authentic, deep, from my toenails to the top of my head; gratitude for all the unbelievable experiences I've had in my life so far. I can't put into words how grateful I am for Dr. Wilson and his brilliance in finally breaking my decades-long self-hate and depression. He freed me from my own mental prison, and every day the sun looks brighter, thanks to him.

Brad thought I should call this book *For All the Girls Who Dream, You'll Find Yourself in You.*

And isn't that the truth? From my little-Dutch-boy haircut to the rock-and-roll hair fiasco in Chicago, at every job I get, they always want to put their mark on me—change my hair or clothing to make me what they think I should be. I always go along with it because that is part of the gig, but within a year, I always come back to the same cut and color, the same style of clothing. I come back to me. It's strange that hair has been a metaphor for life here, but there it is.

EPILOGUE

I know life isn't always going to be this bright and shiny, because I have seen the other side. But now I have the tools to deal with whatever life brings me. I will never allow myself to be mistreated again. I will bask in this sunshine and never let go of my stormy days. I am a natural disaster, and that's okay. And if you are too, congratulations. You can be a natural disaster, but you must do it responsibly. The impulsiveness, the messiness, the constant yearning for bigger and better; those can be good qualities. Embrace them, and treat others with respect, but above all, respect yourself. Saying yes is almost always the right move, but knowing your value and knowing when to say no can take a natural disaster to the next level. Much like the storms I cover, those natural disasters have purpose, and eventually gratitude can be found within each. Now I know how to *be* a natural disaster with gratitude.

If anyone reading this is struggling with depression, please ask for help. We all have people who love us and we all have the ability to raise our hand and say, "I am not okay." It may seem like the worst idea right now, but please do it. You are worth it and this moment will pass. Life is bright and so worth living. Whether you just need a friend to talk to or need to check yourself into a hospital to get a fresh start and some serious therapy, there is no harm nor embarrassment in either. There are so many forms of depression so by no means do I think this covers even the start of mental illness, but it is my story and I am lucky to be here to tell it. This story is so difficult to end, because I know it isn't finished. I have so much more left to become grateful for, and I'm looking forward to it all.

ACKNOWLEDGMENTS

This is the part of the book where I get to thank everyone. Especially the people who have consistently said, "I can't believe I'm not in your book."

Thank you to my mom for always letting me tell your stories, knowing that my heart is full of all love and respect for you. My stepdad, Carl, for keeping our family mentally healthy. My dad, and all my sisters and brothers: Sean, Bridget, and Walter Zuidgeest, and Adrianna and Elaina Craft.

Thank you to Wendy Lefkon and Lisa Alden for guiding me through the publishing process.

Thank you Sam Wnek, my producer of four years, friend of six, and best phone-holding, picture-taking meteorologist out there. Thanks to the team I've had the honor of working with behind the scenes, too—Max Golembo, Melissa Griffin, Dan Peck, and Dan Manzo.

To my lifelong friends Alysha Kirkwood, Liz Neeley, Kelley Iuele, and Lindsey Savickas, for being the best band of friends ever since we were thirteen, and seeing me through all these messes.

To Cindi Poll, my neighbor and first boss at the country club where I learned how to work harder than I ever imagined. Cindi gave me confidence and opportunity to grow my social skills that have carried me throughout life.

To my many mentors, including James Spann, Colleen Pierson, Peter Chan, Tom Skilling, and John Knox.

To James Goldston, current president of ABC News, thank

you for the unbelievable opportunities to travel the world with drones, bringing people to places few have ever seen. So much left to come.

Francesco Bilotto, a friend I met in an odd coincidence on the same day I met Ben. He would from that day on help me turn my sense of design and my physical surroundings from "lesbian dorm room" into "glamorous New York City meteorologist." He claims that transition is responsible for my husband's interest in me, because has he says, "No one wants to marry a little piggy."

Lisa Hayes, Juanita Townsend, and Jamie Salazar, who have been my glam team, keeping me looking sharp even on those mornings I most definitely am not.

My agents Jay Sures, Andrew Lear, Paul Fedorko, and Rick Ramage.

My coworkers behind the scenes and on the air—Michael Strahan, Amy Robach, George Stephanopoulos, Lara Spencer, and especially Robin Roberts, who has been a mentor and the most generous coworker and friend.

My son, Adrian, who I guess will someday read this. I want him to know how he changed my life forever in the best possible way.

To my husband's family—Janis, Mark, Randy, Ernie, Traci, Cameron, Jennifer, and Ella—thank you for accepting me as one of your own.

Most of all, thank you to my husband, Ben Colonomos. I know no other man that could read about his wife's past, enjoy it, and help her edit it.

Enjoy this excerpt from Ginger Zee's new book,
A Little Closer to Home,
available now wherever books are sold.

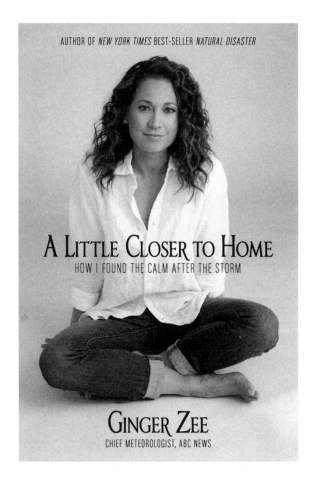

CHAPTER ELEVEN
MY FIRST MEDITATION

Sun-kissed in the early light of a summer Saturday in Oklahoma, Mount Scott and the towering rocks around it were glowing in majestic pinks, reds, and purples. I was in the Wichita Mountains Wildlife Refuge. It's a truly special place. In what is otherwise endless prairie land, this nearly sixty-thousand-acre refuge that escaped destruction because the soil was too rocky to plow, is now home to herds of bison, elk, deer, long-horned cattle, and hundreds of other protected animals. My sense of wonderment was extraordinary.

When I looked down, I noticed something that on any other day would not have even registered: a shallow puddle. I crouched down to get a closer look at the otherwise trivial divot in the land that allowed the recent rain to collect. But it wasn't trivial at all, it was intricate and complex, teeming with life.

Dozens of tiny water bugs were doing their thing, jumping

from side to side, swimming below the surface. They almost appeared to have a pattern they were following. Each one, I thought, had a name, a place in their bustling water-bug village. I sat in awe of the structure these insects were operating under. I sat for so long I even started imagining which little area was their post office or grocery store. The detail was surreal and comforting. I felt so connected to the little bugs because on a grander level, I believe there must be something or someone else looking down at humans on Earth with the same awe.

I would love to tell you that I came to this philosophical moment after years of studying meditation and Buddhism, or at that very moment in my late twenties I was having a breakthrough that would forever benefit my mental health.

Nope. I was hunched over, admiring a small pile of insects for an hour because I had just ingested a mason jarful of fermented cactus. I was on peyote.

Before you roll your eyes and think, *Oh goodness, she wasn't just abusing alcohol, it was drugs, too*, I swear on my children this was the only time I have ever experienced a drug grander than marijuana. And I don't regret it for a minute.

There is such beauty in this world. It's all around us. We just don't stop long enough to see it. You hear that constantly, but for the first time in my life, I was actually living it.

A little background. This was just months before my hospitalization. When a group of friends announced they would be meeting a shaman in the Wichita Mountains of Oklahoma, I said, "Why the heck not?"

I am not certain what I was expecting a shaman to look like. If I am honest, it would have been something more reminiscent of a monk perhaps? Instead, there was a guy who looked pretty much like a grad student who had taken a wrong turn trying to get back to campus. He lit some sage and walked us through what we could expect in the following eight hours.

Again, I had never experimented with drugs aside from a puff of a joint here or there (see, I know that a *puff of a joint* will tell you just how few times I have smoked), so I really had no base to go from. He told us there may be some mild digestive discomfort and then pure bliss. From his earthy-looking suede backpack he unwrapped several mason jars full of a thick dark greenish-brown liquid. He distributed a jar to each of us, waved the sage again, and disappeared after we had all gulped down the thick, mucous cactus guts.

The shaman suggested we hike into the mountains and experience this in nature, because it's not recommended to be in an urban or residential setting when you are about to trip for an entire day. If a pile of water bugs blew your mind, imagine seeing Elmo in Times Square.

I had been to those mountains before, and they were good for an easy hike and to take in some fresh American-plains air with a picturesque backdrop. The beginning of our hike felt very similar to the other times I had ventured into the Wichitas.

We had been hiking for ten minutes when one of the other women with us asked us all to slow down. She was feeling nauseous. She went behind a rock and started vomiting. Her boyfriend clutched at his lower intestines, obviously in pain. Another guy we were with started farting. It was all quite comical but a little scary, not knowing how long this part was going to last.

I just kept thinking, *Oh good, I am one of those that won't be getting the gastro impact.* But before that grateful thought even left my brain, I suddenly felt like someone had punched me in the stomach as hard as they could. I doubled over, crossing my arms and collapsing to the ground with a loud "Owww." I was in pain!!! But as quickly as it came on, it was gone. The gas passed and I was born again.

That's how it felt. I felt like I was being born into a new body and world. This new world was full of brilliant colors, euphoric feelings, and love-filled, blissful people everywhere.

We kept hiking for what felt like an hour before we came to the edge of a dazzling landing overlooking the valley below and more mountains in the distance. The red-clay colors of what is already a glorious landscape were made brighter and more vibrant by the drug. I really wish I could help Instagram create a filter for our pictures that would imitate the power of peyote.

Wow, what filter do you have on there?

Peyote.

Oh, love that one.

I don't think everyone needs to do peyote, but I really would love for all of you to be able to see and feel the world like I did that day. My mind was open.

The six of us sat independently, dozens of feet apart, yet we felt like we were holding hands and experiencing it all together. I got very focused on the mountains in the distance. While staring at them for what I am sure was thirty minutes but felt like three, I could see history, not just rock, jutting out from the land. There were three descending heights of rocks and as I stared, I realized they were telling me a story. The rocks looked undeniably like a horse with a man riding it and another man towering over them both with a gun and a sword. I could see the fear and strain in the horse's face. Tears started rolling down my cheeks and I began praying. It was so moving, and while I had known that story my entire life, it was touching me from the very ground that I sat on, because the peyote was allowing me to stay still, feel, and know it in my core.

From that spot, someone in the group got up enough motivation to keep hiking, so the rest of us followed. The day was perfect: low 70s, sunshine. We walked and stopped every few minutes. One of those stops was the water bugs, others were just long, deep breaths while watching the blue sky. One of the last stops, as the sun was setting, took us to the valley floor where dozens of trees stood, bare of leaves and surrounded by blackened sand. It was obvious a fire had recently blown through this part of the refuge. As I got in to take a closer look at the swirling black sands and dirt that encircled the trees, the landscape again transformed. This time it wasn't history but an art reference, the scene in front of me morphing into a Salvador Dalí creation. The twiglike trees became more cartoonish, like his tree in *The Persistence of Memory*. The surrealist painting came to life and I was walking through the oils, touching them and moving them like I was an artist within the art.

My invisible paintbrush started to lose its power, and that's when I knew my mind was closing and the drug was wearing off. We all started hiking out of the burned field of trees because it was almost dark and no longer safe for us to be tripping around. At dusk we were met again by the shaman who had stood watch over us the entire day (part of the package deal). We all started feeling our old selves again but were forever changed by the experience.

Fast-forward from that day to a few years later, when I met Dan Harris at ABC News. He had just written his life-changing book, *10% Happier* (coincidentally born from experimenting with cocaine, which led to a panic attack and then a deep dive into meditation). Dan introduced me to meditation as he has done for countless others. Sure, I had heard of it, but he explained it to me like no one else had. It is like a treadmill for our mind. It will make us stronger. Stronger in mind is exactly what I needed in my life at that moment. I was going to therapy and finally learning my emotional-regulation tools, but this was another step I could do at home.

I read his book and gave it my all. I would sit, quietly, and try my best not to fall asleep (meditation may be more difficult for narcoleptics as a side note). Dan is a great coach. He kept assuring me that meditation is not about *thoughtlessness* but *thoughtful thoughts*. Our *monkey brains* as he calls them will forever be on the cycle of ping-ponging and running full speed, but meditation is the strength in allowing those thoughts to be just what they are: thoughts. Not acting on them, not writing them down. Just saying, *Okay, Ginger, that's a nice thought about how you need to look at peel-and-stick wallpaper, and yes, a nice thought about how Miles's bike helmet is getting too small. Nice thoughts. Nothing to do with them now. Let them go. They're just thoughts—let them go.*

You start to notice patterns about your thoughts and can start to almost anticipate them. This gives you control in your everyday life and not just that quiet moment on your cushion. For many people this settles them down. Their thoughts would often sweep them into a fury of anger or frustration. For me, meditation taught me how to drop in and feel. Settling and observing and not ignoring what I feel, thinking about what's happening around me. I really believe there is power in meditation, and I wrote a whole different chapter about this later in the book, about how it helped me work on my emotional regulation.

I certainly can't do peyote every day, but in some of my most meditative moments I have felt a glimmer of the same euphoria. My mind has been open enough, and I have the Wichita Mountains to aspire to.

The best part? After all the stories were told that night back in Oklahoma as we drove home, the shaman asked, "How far do you think you hiked?" We all guessed two to ten miles. I thought it had to be somewhere close to six miles at least. He informed us we had barely made it in a quarter-mile loop outside the parking lot. All that beauty, all that open-mindedness is right in front of us. That's what I never let go from my experience that day.

With our blinders on, we run full-steam ahead doing twenty miles only to have not seen a damn thing along the way. Sometimes you can do less to see more. Just take the blinders off—there are some water bugs that need your attention.